108 Advances in Polymer Science

W0232244

Structure in Polymers with Special Properties

Editor: H.-G. Zachmann

With contributions by
F. J. Baltá Calleja, F. Batallán, T. Ezquerra,
B. Frick, A. González Arche, H. G. Kilian,
E. López Cabarcos, K. Miyasaka, T. Pieper,
C. Santa Cruz

With 92 Figures and 5 Tables

 Springer-Verlag Berlin Heidelberg GmbH

Guest Editor:

Prof. Dr. H.-G. Zachmann
Institut für Technische
und Makromolekulare Chemie
Universität Hamburg
Bundesstr. 45, 20146 Hamburg, FRG

ISBN 978-3-662-14941-6   ISBN 978-3-540-47594-1 (eBook)
DOI 10.1007/978-3-540-47594-1

This work is subject to copyright. All rights are reserved, whether the whole or part
of the material is concerned, specifically the rights of translation, reprinting, re-use
of illustrations, recitation, broadcasting, reproduction on microfilms or in other ways,
and storage in data banks. Duplication of this publication or parts thereof is only
permitted under the provisions of the German Copyright Law of September 9, 1965,
in its current version, and a copyright fee must always be paid.
© Springer-Verlag Berlin Heidelberg 1993
Originally published by Springer-Verlag Berlin Heidelberg New York in 1993.
Softcover reprint of the hardcover 1st edition 1993

Library of Congress Catalog Card Number 61-642

The use of registered names, trademarks, etc. in this publication does not imply, even
in the absence of a specific statement, that such names are exempt from the relevant
protective laws and regulations and therefore free for general use.

Typesetting: Macmillan India Ltd., Bangalore-25

02/3020 5 4 3 2 1 0 Printed on acid-free paper

Preface

Following a period during which huge amounts of conventional plastics, films and fibers were produced for application in everyday life, polymers with special properties are becoming increasingly important. High performance polymers, photopolymers and ferroelectric polymers are some of the materials on which science and technology is focussed at the present time. As with all polymers, the properties of these materials are strongly related to their structure, particularly to the molecular order and orientation and to the supramolecular structure. Therefore, it seems important to summarize and discuss some of the new results in this field of research.

In the article by Baltá Calleja et al., the latest results of investigations into the structure of poly(vinylidene fluoride) and its copolymers with poly(trifluoroethylene) are summarized and extensively dicussed. These polymers are the most important ferroelectric materials. Special emphasis is placed on the relation between the change of structure and the transition from the ferroelectric into the paraelectric phase.

In the contribution written by Miyasaka the complex of system poly(vinyl alcohol) and iodine is discussed. Such systems are important materials as film polarizers. It is shown that different structures can be formed depending on the concentration of iodine and the degree of orientation of the polymer. The optical behaviour is explained by the structure of the material.

A general concept for describing all kinds of order in chain molecules, ranging from crystalline order to liquid crystalline order and then to order in oriented and isotropic amorphous polymers, is introduced in the third article written by Pieper and Kilian. After the presentation of the basic concept, experimental results obtained on different polymers including phases with rotatory segmental motion are discussed.

The publisher and the editor hope that these articles written by well known experts in their fields of research will be of benefit to all scientists working on the structure and properties of polymers.

Hamburg, March 1993   H. G. Zachmann

Editors

Prof. Akihiro Abe, Tokyo Institute of Technology, Faculty of Engineering, Department of Polymer Chemistry, O-okayama, Meguro-ku, Tokyo 152, Japan

Prof. Henri Benoit, CNRS, Centre de Recherches sur les Macromolécules, 6, rue Boussingault, 67083 Strasbourg Cedex, France

Prof. Hans-Joachim Cantow, Institut für Makromolekulare Chemie der Universität, Stefan Meier-Str. 31, 79104 Freiburg i. Br., FRG

Prof. Paolo Corradini, Università di Napoli, Dipartimento di Chimica, Via Mezzocannone 4, 80134 Napoli, Italy

Prof. Karel Dušek, Institute of Macromolecular Chemistry, Czechoslovak Academy of Sciences, 16206 Prague 616, TSEC

Prof. Sam Edwards, University of Cambridge, Department of Physics, Cavendish Laboratory, Madingley Road, Cambridge CB3 OHE, UK

Prof. Hiroshi Fujita, 35 Shimotakedono-cho, Shichiku, Kita-ku, Kyoto 603 Japan

Prof. Gottfried Glöckner, Technische Universität Dresden, Sektion Chemie, Mommsenstr. 13, 01069 Dresden, FRG

Prof. Dr. Hartwig Höcker, Lehrstuhl für Textilchemie und Makromolekulare Chemie, RWTH Aachen, Veltmanplatz 8, 52062 Aachen, FRG

Prof. Hans-Heinrich Hörhold, Friedrich-Schiller-Universität Jena, Institut für Organische und Makromolekulare Chemie, Lehrstuhl Organische Polymerchemie, Humboldtstr. 10, 07743 Jena, FRG

Prof. Hans-Henning Kausch, Laboratoire de Polymères, Ecole Polytechnique Fédérale de Lausanne, MX-D, 1015 Lausanne, Switzerland

Prof. Joseph P. Kennedy, Institute of Polymer Science, The University of Akron, Akron, Ohio 44 325, USA

Prof. Jack L. Koenig, Department of Macromolecular Science, Case Western Reserve University, School of Engineering, Cleveland, OH 44106, USA

Prof. Anthony Ledwith, Pilkington Brothers plc. R & D Laboratories, Lathom Ormskirk, Lancashire L40 SUF, UK.

Prof. J. E. McGrath, Polymer Materials and Interfaces Laboratory, Virginia Polytechnic and State University Blacksburg, Virginia 24061, USA

Prof. Lucien Monnerie, Ecole Superieure de Physique et de Chimie Industrielles, Laboratoire de Physico-Chimie, Structurale et Macromoléculaire 10, rue Vauquelin, 75231 Paris Cedex 05, France

Prof. Seizo Okamura, No. 24, Minamigoshi-Machi Okazaki, Sakyo-Ku, Kyoto 606, Japan

Prof. Charles G. Overberger, Department of Chemistry, The University of Michigan, Ann Arbor, Michigan 48109, USA

Prof. Helmut Ringsdorf, Institut får Organische Chemie, Johannes-Gutenberg-Universität, J.-J.-Becher Weg 18-20, 55128 Mainz, FRG

Prof. Takeo Saegusa, KRI International, Inc. Kyoto Research Park 17, Chudoji Minamima-chi, Shimogyo-ku Kyoto 600 Japan

Prof. J. C. Salamone, University of Lowell, Department of Chemistry, College of Pure and Applied Science, One University Avenue, Lowell, MA 01854, USA

Prof. John L. Schrag, University of Wisconsin, Department of Chemistry, 1101 University Avenue. Madison, Wisconsin 53706, USA

Prof. G. Wegner, Max-Planck-Institut für Polymerforschung, Ackermannweg 10, Postfach 3148, 55128 Mainz, FRG

Table of Contents

Structure and Properties of Ferroelectric Copolymers of Poly(Vinylidene Fluoride)

F.J. Baltá Calleja[1], A. González Arche[2], T. A. Ezquerra[1],
C. Santa Cruz[1], F. Batallán[3], B. Frick[4], and E. López Cabarcos[5]
[1] Instituto de Estructura de la Materia, CSIC, Serrano 119, Madrid 28006, Spain;
[2] EUIT, Univ, Politécnica Madrid;
[3] Inst. de Ciencia de Materiales CSIC, Madrid;
[4] ILL, Grenoble;
[5] Fac. Farmacia, Univ. Complutense Madrid

The aim of this paper is to review the main structural features and properties which are related to the ferroelectric behavior of copolymers of poly(vinylidene fluoride) (PVF_2). These are illustrated by describing the conformation, polymorphism and morphology of PVF_2 and its copolymers. Copolymers of PVF_2 and trifluoroethylene have special interest because of their prominent ferroelectric behavior as well as strong piezo- and pyroelectric activity. The review covers the study of the influence of molar composition on structure and the temperature induced ferroelectric to paraelectric phase changes. Advances and problems in X-ray diffraction results relating to changes in the long period, size of coherently diffracting domains and lattice spacings with temperature are discussed. A microstructural model to explain the transformation of paraelectric into ferroelectric crystals when cooling through the transition temperature is presented. Special attention is paid to the influence which the changes in structure exert upon mechanical properties. The dependence of polarization upon copolymer composition and temperature is analyzed. Dielectric spectroscopy is shown to be of great interest in showing transitions connected with reorientation of permanent dipoles. Finally, new results on molecular dynamics of these ferroelectric copolymers derived from incoherent neutron scattering are reported.

Advances in Polymer Science, Vol. 108
© Springer-Verlag Berlin Heidelberg 1993

1 Introduction

In recent years there has been increasing recognition of the importance of polymeric materials which can be incorporated as active elements in electric circuits [1]. Most of these materials are piezoelectric. Owing to the complex molecular, crystalline and morphological structure of polymers, it appears, nevertheless, surprising that there might exist any polymer which complies with the restrictive requirements of piezoelectricity and its two related properties: pyroelectricity and ferroelectricity. A substantial piezoelectric effect was not found in synthetic polymers until the investigations of Kawai in 1969 [2] on elongated and polarized films. In this work the largest effect was exhibited by poly(vinylidene fluoride) (PVF_2) whose molecular repeat formula is (CH_2–CF_2). Since this time, much attention has been focused in many laboratories on this material and on other polymer systems having a large dipole moment in their chemical repeat unit. The presence of pyroelectricity was reported in 1971 [3]. Whether PVF_2 was a true ferroelectric material rather than a trapped charged electret was a controversial issue for about a decade after the discovery of its strong piezoelectricity [4]. Dipolar reorientation during application of a high electric field has been proved by X-ray techniques [5].

Ferroelectricity has also been found in certain copolymer compositions of VF_2 with trifluoroethylene, F_3E, [6–11] and tetrafluoroethylene, F_4E, [12–15] and in nylon 11 [16]. Specifically, copolymers of vinylidene fluoride and trifluoroethylene (VF_2/F_3E) are materials of great interest because of their outstanding ferroelectricity [9, 17–18], together with a parallel strong piezo- [7] and pyroelectricity [19]. These copolymers exhibit, in addition, an important aspect of ferroelectricity that so far has not been demonstrated in PVF_2: the existence of a Curie temperature at which the crystals undergo reversibly a ferroelectric to a paraelectric phase transition in a wide range of compositions [9, 17–18].

Although piezoelectricity in synthetic polymers is not as high as in natural occurring single crystals and ceramics, the advantage of these polymeric materials as piezoelectric elements arises because they have low density, are more flexible, and easier to process than conventional ceramics and single crystals. They also show a much higher dielectric strength than ceramics together with a lower mechanical and acoustic impedance which makes them good candidates as sensors of mechanical signals. PVF_2 is already being used in many applications such as transducers in ultrasonic cardiac imaging, blood pressure and pulse measurements, touch sensors in robotics [20], optical fiber coatings for electric field sensing [21] and, recently, as a programmable neural network system [22]. Piezoelectric polymers are also being used in composites with piezoelectric ceramics to improve their properties in device applications [23].

In this review the basic concepts of piezoelectricity, pyroelectricity and ferroelectricity will be first considered in the light of molecular structure. The salient features of the molecular conformations of PVF_2 and its copolymers in the crystalline state will be presented. This will be followed by a comprehensive account of the chain packing possibilities for these molecules with a regular conformation giving rise to various polymorphic forms. This leads directly to the problem of relating the various crystalline modifications to the different spherulitic morphologies occurring in melt-solidified PVF_2. In addition to the wide morphological diversity various types of crystalline transformations will be discussed. The structural changes in the ferro-paraelectric phase transition of the vinylidene fluoride/trifluoroethylene copolymers with various compositions will be discussed in the light of wide and small angle X-ray scattering and DSC experiments. The temperature dependence and effect of electron irradiation upon the mechanical properties (microhardness, young modulus) of these ferroelectric copolymers will be presented. Furthermore, the morphological changes occurring at the ferro-to-paraelectric transition which is accompanied by rotational motion of chains will be correlated to the dielectric and ferroelectric behaviour of these materials. The present review will be completed by presenting novel molecular dynamic data of these ferroelectric copolymers using incoherent elastic and quasielastic neutron scattering and the analysis of the various motions in the various phases.

2 Basic Concepts: Piezo-, Pyro- and Ferroelectricity

A piezoelectric material is one that undergoes a change in electron polarization in response to a mechanical stress or vice versa. However, only certain molecular crystals without a center of symmetry produce an electric charge when mechanically deformed. This phenomenon is attributed to the net internal polarization in the crystal. When no external forces are present, the centers of positive and negative charges will coincide with each other and there is no net polarization. The application of a stress, be it mechanical (pressure) or electrical (applied field) causes a displacement of the centers of gravity of positive and negative charges. In the absence of a center of symmetry the charge displacement will be non-symmetrical and, thereby, produces an induced dipole moment. This dipole moment, if produced by a mechanical stress will cause the surface to develop an effective charge. If an external field displaces the charges by electrostatic attraction or repulsion it produces the mechanical strain which causes the material to deform. The special characteristic of the piezoelectric effect is, thus, a true electro-mechanical transducer action which varies in a first approximation linearly with the electric field and mechanical stress [24]. This implies that the effect operates on permanent electric dipole moments and alters the balance of such moments.

Of the 32 crystal classes, 20 lack a center of symmetry. Materials built up by these crystals have, therefore, the potential for piezoelectric activity.

The most useful piezoelectric constant is the tensor d_{ij} which relates electric polarization to the stress causing the polarisation. The d-constant is also identified to the derivative of the resulting strain with respect to the applied electric field [24]:

$$d_{ij} = (\delta P_i / \delta \sigma_j)_E = (\delta \varepsilon_i / \delta E_j)_\sigma \tag{1}$$

where the first subscript i designates the electrical direction (electric polarisation) and the second subscript j identifies the mechanical direction. The piezoelectric strain constant d is easy to measure and is a property which can be readily incorporated into mechano-electrical and electro-mechanical devices [25].

Among the 20 crystal classes lacking a center of symmetry ten of them contain a unique polar axis and exhibit pyroelectricity in addition to piezoelectricity, i.e. in the unstrained dipolar network of these crystals the dipole moment components remain and add to a resultant polar-axis moment. The term pyroelectricity is assigned because thermal expansion will expand or contract the dipole. The pyroelectric constant is defined by:

$$p = (\delta P / \delta T)_{E,\sigma} \tag{2}$$

The length of the polar axis, and with it the polarization, varies with temperature; hence crystals with a polar axis develop, as the temperature changes, a difference in potential: i.e. a pyroelectricity of opposite sign for heating and cooling [26].

Finally, ferroelectricity is manifest in asymmetrical crystals producing domains of spontaneous polarization whose polar axis direction can be reversed in an electric field directed opposite the total dipole moment of the lattice. The two (or more) directions can coexist in a crystal as domain structures comprising millions of unit cells which contain the same electric orientation. The symmetry elements are temperature sensitive in ferroelectric materials [27]. At a particular temperature called the Curie Point the values of the piezoelectric coefficients reach particularly high values. Above the Curie Point the crystal transformation is to a less polar form and the ferroelectric nature disappears.

3 Conformation of PVF$_2$ and Its Copolymers in the Crystalline Phases

As pointed out above, the occurrence of a pronounced piezoelectric effect in synthetic polymers can be mainly attributed to the presence of a non-centrosymmetric unit cell and a net polarization in the material. To achieve this in polymeric materials one requires [28]: (a) a large dipole moment in the chemical repeat unit; (b) feasibility of crystallizing in a non-centrosymmetric unit cell; (c) alignment of molecular dipoles. A dipole moment arises from a charge separation between adjacent atoms. In polymers this can arise between covalently bonded atoms along the polymer chain or between adjacent chains. A significant dipole moment occurs when chlorine or fluorine atoms are incorporated in the carbon backbone. For instance the fluorine atom, which has a van der Waals radius only slightly larger than that of hydrogen, forms highly polar bonds with carbon having a dipole moment of 1.9 Debye (1 Debye $= 3.34 \times 10^{-30}$ C.m). The presence of head-to-head, $(CH_2–CH_2)$, or tail-to-tail, $(CF_2–CF_2)$, defects in polyfluorocarbons like PVF_2 is found to reduce the average dipole moment of the chain segment $(CH_2–CF_2)$ by 6–10% [29]. For poly(vinyl fluoride) [PVF, molecular formula $(CH_2–CHF)_n$] these types of defects are more common than in PVF_2 and lead to reductions in the dipole moment of 20–24%. For polytrifluoro-ethylene [PF_3E, $(CF_2–CHF)_n$] the lack of isoregularity induces reductions in the dipole moment of 22–26% [30]. Another type of configurational defect which affects the magnitude of the polarization along a chain segment is the tacticity of the polar groups. For PVF the largest dipole moment would occur if all the fluorine atoms were on the same side of the carbon–carbon plane (isotactic conformation). However, owing to the random (atactic) arrangement of the fluorine atoms along the main chain it has a net dipole moment of about 20% of that of ideal isotactic PVF_2 [28].

The molecular chains of most polymers within a crystal lattice are restricted by steric and electrostatic intramolecular interactions into a regular conformation of lowest potential energy. In this respect PVF_2 represents an outstanding exception, as it can adopt three regular conformations of similar potential energies for reasons associated with the van der Waals radii of its constituents. Excellent reviews describing the structural requirements for ferroelectric behavior in PVF_2 are available [28, 31, 32]. Therefore, in what follows we shall just try to offer a condensed view of these aspects.

PVF_2 molecules containing two hydrogen and two fluorine atoms per repeat unit, are intermediate between chain molecules rich in hydrogen like PE $(CH_2–CH_2)$ and chain molecules rich in fluorine like PF_4 $(CF_2–CF_2)$. In PE only low rotational barriers separating its possible conformations are found. Therefore, only the lowest energy conformation (all-trans) is reached in this case. In PF_4, rotation is sterically hindered so-that these molecules are forced to adopt one conformation which is normally helical. PVF_2 molecules, on the contrary, have a choice of multiple conformations like the hydrogen rich macromolecules. However, since rotational barriers in this case are high, chains can also be stabilized into favorable conformations other than that of lowest energy. The known conformations of PVF_2 molecules in the crystalline phase are the following ones: all-trans; the tg^+tg^- and the $tttg^+tttg^-$. In all cases there are slight deviations from the 180° and $\pm$ 60° torsional angles. The first two conformations are the most common and important ones and are shown in Fig. 1. Owing to the alignment of all its dipoles in the same direction normal to the chain axis the all-trans is the most highly polar conformation in PVF_2 (2.1 Debye per repeat). The tg^+tg^- conformation is also polar but because of the inclination of dipoles to the molecular axis (Fig. 1) it has components of the

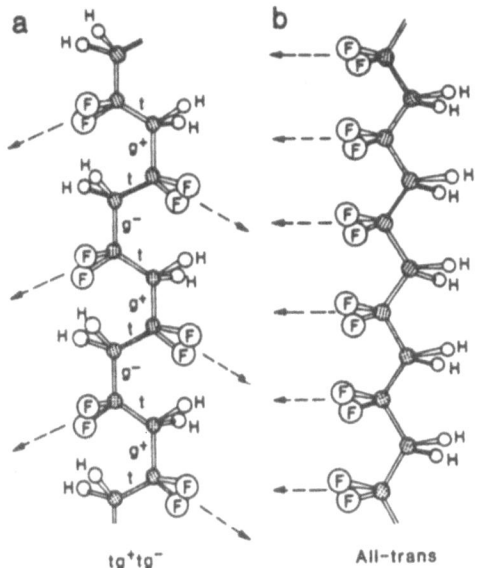

Fig. 1a, b. Schematic depiction of the two most common crystalline chain conformations in PVF$_2$: **a.** tg$^+$ tg$^-$ and **b.** all-trans. The *arrows* indicate projections of the –CF$_2$ dipole directions on planes defined by the carbon backbone. (Figure and caption from Ref. [32])

dipole moment both perpendicular (1.2 Debye) and parallel (1 Debye) to the chain. Approximately the same value characterizes the tttg$^+$tttg$^-$ conformation.

On the other extreme some studies were performed on pure poly(trifluoroethylene), PF$_3$E, [33–36]. This polymer is essentially atactic, nevertheless it crystallizes, because the fluorine and hydrogen atoms behave isomorphically allowing development of significant crystallinity. The X-ray diffraction pattern of PF$_3$E is very poor and the few reflections observed can be indexed with an hexagonal lattice having a = b = 5.6 Å. Concerning the c parameter, values of 2.25 Å [33], 2.29 Å [34, 35] and 2.50 Å [36] have been reported. According to Lando et al. the conformation of PF$_3$E is of the 3/1 helix type [33]. However, according to Lovinger [35], the diffuse meridional reflection at 2.29 Å is consistent with an irregular succession of tg$^+$, tg$^-$ and tt groups, while Tashiro [34] proposes a disordered all-trans conformation. It seems apparent that molecules having less than 20% head-to-head defects should adopt a 3/1 helical conformation while those having a higher defect content should favor a trans arrangement.

Copolymers of VF$_2$ with trifluoroethylene are randomly added copolymers. Those containing a mole fraction of VF$_2$ of 50–80% have been widely studied. Since they contain a greater proportion of the comparatively bulky fluorine atoms than PVF$_2$ their molecular chains cannot accommodate the tg$^+$tg$^-$ conformation and crystallize at room temperature in the ferroelectric phase with the extended all-trans planar conformation [37] with small statistical deviations away from that plane, i.e. copolymers of VF$_2$ with F$_3$E crystallize essentially with the same conformation as β-PVF$_2$.

Copolymers of VF$_2$ and trifluoroethylene also exhibit a Curie temperature at which the ferroelectric crystals show reversibly a solid state transformation to

a non-polar paraelectric state (see Sect. 7) [7, 18, 38–39]. The Curie transition
has been found to involve intramolecular changes of dipole directions through
introduction of $g\pm$ bonds that modify the polar all-trans conformation to a
somewhat disordered arrangement of tg^+, tg^- and tt sequences [39].

4 Chain Packing in the Crystals: Polymorphism

Once a polymer molecule has crystallized, it is the presence of a non-centrosymmetric unit cell which gives rise to a resultant dipole moment and associated
piezoelectric activity. The capability of molecules of PVF_2 and related polymers
to adopt polar conformations is, however, not sufficient to ensure polarity of
their resulting crystals because the molecules may be packed in a lattice in such a
way as to cancel their dipole moments. For instance, PVF_2 in the α-form
crystallizes with two chains per unit cell having the tg^+tg^- conformation
(Fig. 2a) with dipole components normal to the chain axes which are, however,
antiparallel. In this type of packing the dipoles of neighbouring chains neutralize
each other [39]. The α phase is the most common polymorph of PVF_2 and can
be easily obtained by melt crystallization at any temperature [39]. Experimental
evidence supports the view that the disposition of axial components of the

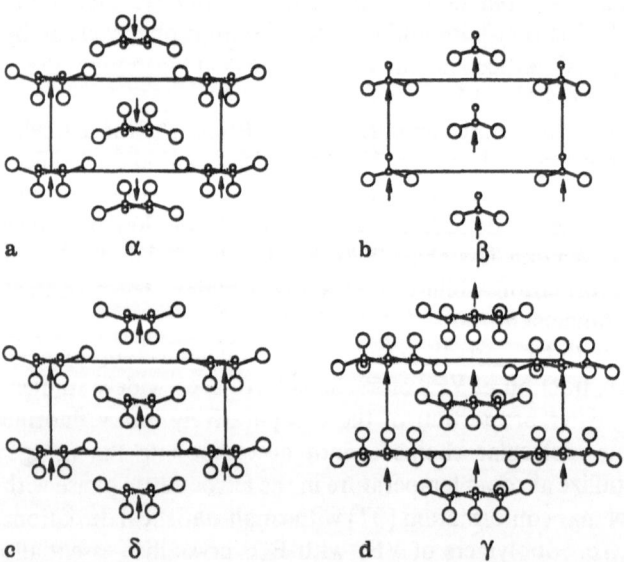

Fig. 2. Projection of four crystal structures of PVF_2 viewed along the molecular axes. Dipole
moments are shown by *arrows*. *Large circles* represent fluorine, *smaller circles* represent carbon.
Hydrogen atoms are not shown. (Figure and caption from Ref. [55])

dipole moments for the α-chains are generally in the form of a statistical packing which becomes regularly antiparallel upon heat treatment at high temperatures.

By poling with a high voltage the α-form transforms, without any appreciable change in the unit cell dimensions, into a new crystal form δ in which all dipole moments become parallel [40]. In fact, this transformation involves a rotation of every second chain by 180° about its axis so that molecules are now packed with the transverse components of their dipoles moments pointing in the same direction (Fig. 2c). Dipole reversal upon poling may involve propagation of a twist wave along the chains [41] or small intramolecular conformational rotations [42].

The highest piezoelectric response is associated with the polar β-form (Fig. 2b) which exhibits the maximum dipole moment. This pseudo-hexagonal unit cell contains two all-trans chains packed with their dipoles parallel to the b-axis [39]. Packing of chains in β-PVF_2 is such that fluorine and hydrogen atoms of neighbouring chains are approximately at the same level parallel to the a-axis of the unit cell. This favorable intermolecular-contribution to the molecular energy is very important in stabilizing the crystalline structure of β-PVF_2.

In the γ-phase of PVF_2 shown in Fig. 2d, chains of the $tttg^+tttg^-$ conformation are packed with the dipole moments parallel within the unit cell yielding a polar phase [45, 46]. The latter has the same base dimensions as that of its α-counterpart, a fact that allows a solid state transformation from α-PVF_2 to the thermodynamically more stable γ-phase at high temperatures ($\sim 160\,°C$) through limited intermolecular motions [47]. This variety of crystalline phases found in PVF_2 is unusual among polymers – one or two phases are common – and is another result of its unique molecular structure. PVF, on the other hand, crystallizes with only one unit cell that is practically the same as that of β-PVF_2.

At room temperature the unit cell for the VF_2/F_3E copolymers with a VF_2 mole fraction of less than 82% is essentially the same as that of β-PVF_2. The ab cell projection of one representative composition of these copolymers (mole fractions of 73%/27% VF_2/F_3E) and of β-PVF_2 is depicted in Fig. 3. We can see that in both cases the molecular conformations are the same. The intermolecular lattice structure, chain packing and dipolar alignment are also the same. The main structure difference between the homopolymer and its copolymers involves the intermolecular spacing of their lattice. As can be seen in Fig. 3, both, a and b axes of the unit cell are significantly expanded in the copolymers because of the presence of additional fluorine atoms in tri and tetrafluoroethylene. Since trifluoroethylene is added stereoirregularly [43], there is an equal distribution of fluorine atoms on both sides of the planar zig-zag in VF_2/F_3E copolymers. As result of this lattice expansion the VF_2 units in the copolymers will not be in close contact, contrary to the situation in β-PVF_2 homopolymer. For the β-phase of PVF_2 and its copolymers the pseudohexagonal character of its lattice plays a major role in determining the ferroelectric instead of a pyroelectric character of the material whose dipole directions are reversible rather than stable [44].

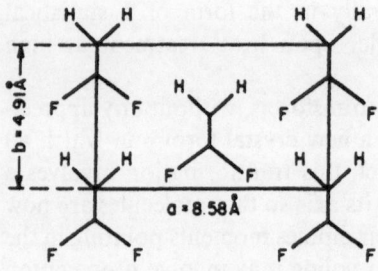

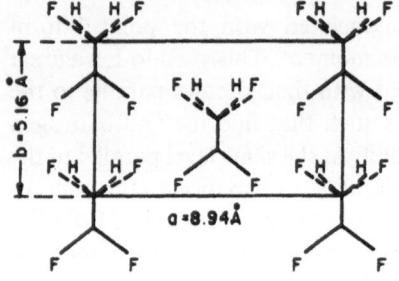

Fig. 3. Unit cell of β-PVF$_2$ (*top*) and of the ferro-electric phase of a VF$_2$/F$_3$E copolymer with mole fractions of 73/27 (bottom), projected along the molecular axis (Figure and caption from Ref. [65])

5 Morphology

We have seen that polarization of individual VF$_2$ repeating units within the chains can prevail through specific levels of structure to give polar crystallites. An area of much importance involves the crystallization of VF$_2$ from the melt and the analysis of the resulting spherulitic morphologies. It is known that PVF$_2$ solidifies from the melt in the form of crystalline lamellar stacks having periodicities of about 10 nm, arranged into spherulites that have no net polarization. Nakagawa and Ishida [48] crystallized PVF$_2$ at high temperatures and studied the distribution of lamellar thicknesses within the spherulites.

A number of authors [49–51] have investigated the morphology of melt solidified PVF$_2$. Two spherulitic types are found to grow from the melt at high temperatures (up to 160 °C). The most common one of the two are the spherulites of the non-polar phase (α-form) which are characterized by their large sizes, high birefringence and tight spaced concentric banding. The spherulites of the second type are, generally, smaller and less birefringent than those of the α-form and exhibit some disorganization and imperfection of crystallographic features as revealed by electron and X-ray diffraction analysis [51, 52]. As a result of this and of the disturbed morphological and birefringent appearance of some of them these spherulites have been called "mixed". Lovinger et al. [51, 52] identified these spherulites predominantly with the γ-form. In addition to these two spherulitic types a third phase is obtained at higher temperatures (T ≥ 160 °C) in melt crystallized PVF_2 owing to a selective solid state transformation occurring in parts of the α-spherulites. Such transformations lead to a form which has been associated with the γ-phase [49, 50]. A detailed morphological study [53] has shown that the solid state transformation in spherulites of α-PVF_2 is initiated in areas of their peripheries where the lamellae are intermeshed with those of mixed spherulites [53]. Typical spherulites of α-PVF_2 grown in our laboratory at 156 °C for 12 hours from the molten state, containing spherulites of the second form are shown in Fig. 4. The β-phase is not usually produced from the melt since that requires high pressures [54] or epitaxial techniques [31] but it can be obtained by mechanical deformation or electrical poling [55]. In PVF_2, crystalline lamellae within the spherulites represent about 50 percent of the total mass, the other half being amorphous.

Random copolymers of VF_2/F_3E when crystallized from the molten state above the Curie temperature show a microstructure in the form of very thin needle-like morphological units which are probably semicrystalline. Figure 5a illustrates the needle-like microstructure of the copolymer 80/20 melt crystallized in the paraelectric phase observed at 140 °C. After cooling at room temperature the microstructure of the ferroelectric crystals is such that what appear in the optical microscope as radial fibers are, in fact, stacks of thin platelet-like morphological units (see Fig. 5b).

In summary, while some of the above polymers can form polar crystals, their three-dimensional arrangement in macroscopic specimens results in internal electrical compensation. To overcome this, such specimens must be externally polarized [55].

Fig. 4. Typical appearance of α-PVF_2 spherulites. Sample cast from dimethyl formamide, molten at 200 °C and recrystallized at 156 °C for 12 hours. Scale bar, 25 μm

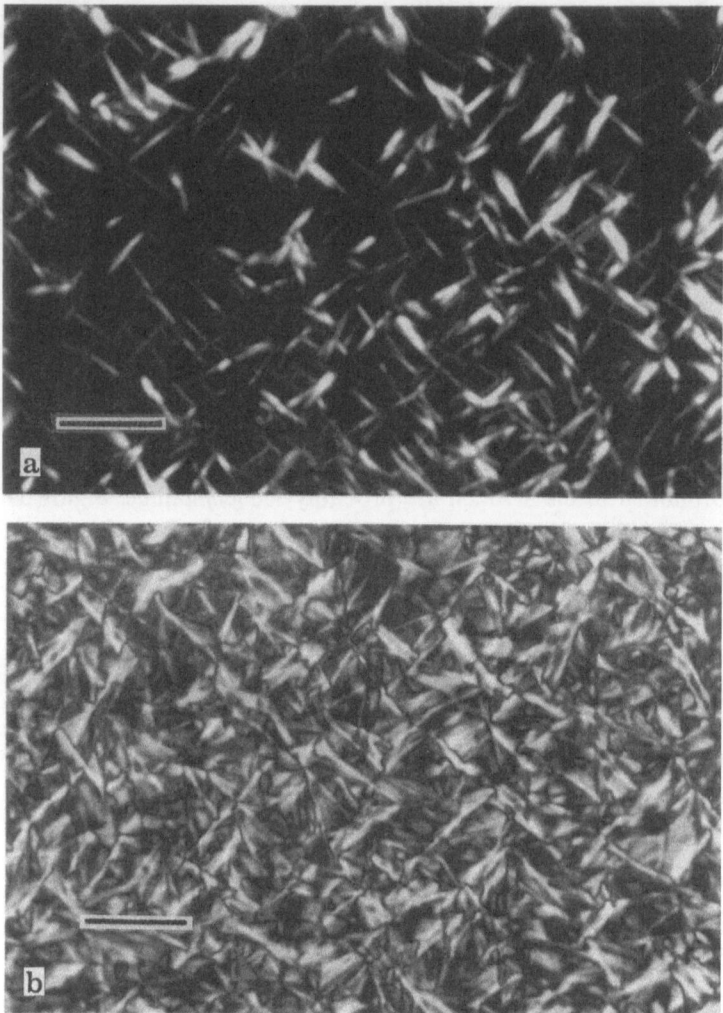

Fig. 5. a. Needle-like microstructure of the 80/20 copolymer. Sample cast from dimethyl formamide molten at 180 °C and recrystallized at 140 °C in the paraelectric phase. **b.** Stacks of thin platelet-like crystals of the same copolymer after cooling the sample at room temperature in the ferroelectric phase. Scale bars, 25 μm

6 Influence of Molar Composition on Structure

6.1 Main Features

The properties and structure of VF_2/F_3E copolymers were initially studied by Lando et al. [33, 37, 56] and Yagi et al. [6, 7, 43]. The interest for these materials experienced a remarkable increase when Furukawa et al. [8, 17, 57] and

Yamada et al. [9, 10] demonstrated that the copolymers were ferroelectric over a wide range of molar composition and that, at room temperature, they could be poled with an electric field much more readily than the PVF_2 homopolymer. The main points highlighting the ferroelectric character of these materials can be summarized as follows: (a) At a certain temperature, that depends on the copolymer composition, they present a solid–solid crystal phase transition. The crystalline lattice spacings change steeply near the transition point. (b) The relationship between the electric susceptibility ε and temperature fits well the Curie-Weiss equation. (c) The remanent polarization of the poled samples reduces to zero at the transition temperature (Curie temperature, T_c). (d) The volume fraction of ferroelectric crystals is directly proportional to the remanent polarization. (e) The critical behavior for the dielectric relaxation is observed at T_c.

At room temperature the copolymers may be classified into three groups according to the mole fraction of VF_2 content:

	Percent VF_2
Group A,	$\leq 12\%$
Group B,	12–82%
Group C	$\geq 82\%$

Copolymers of group A are expected to behave similarly to PF_3E. As far as we know no structural data have been reported in this region. The copolymer of lowest VF_2 content for which the Curie transition was observed is the 13/87 [7, 34]. Those in group C, crystallize in a form analogous to that of the α-PVF_2 and do not present a Curie transition. Copolymers of group B present specific features and properties which are different from those of the corresponding homopolymers. When copolymers in group B are melt crystallized at room temperature they present a mixture of two phases: a well ordered ferroelectric phase, F_F, [31, 32, 34, 38, 39, 58–61] (see for example Fig. 3) and a less ordered phase with non-polar character that we shall call F_{NF}. In copolymers of group B the content of the ferroelectric phase increases with VF_2 concentration. For mole fractions of VF_2 higher than 60–65%, the F_F phase is normally the only one present. However, under special sample preparation procedures, such as quenching from the melt to room temperature, a small fraction of the F_{NF} crystals can also coexist with the polar phase [62]. Copolymers in group B present D–E (electric displacement– electrical field) hysteresis loops that, at high VF_2 content, are nearly square shaped with a large remanent polarization.

6.2 Wide Angle X-ray Diffraction Studies

The X-ray diffraction patterns of the copolymers in group B measured at room temperature present four or five reflections that can be indexed in an ortho-rhombic unit cell (pseudohexagonal sometimes called orthohexagonal). Figure 6 illustrates a typical pattern for the copolymer 80/20.

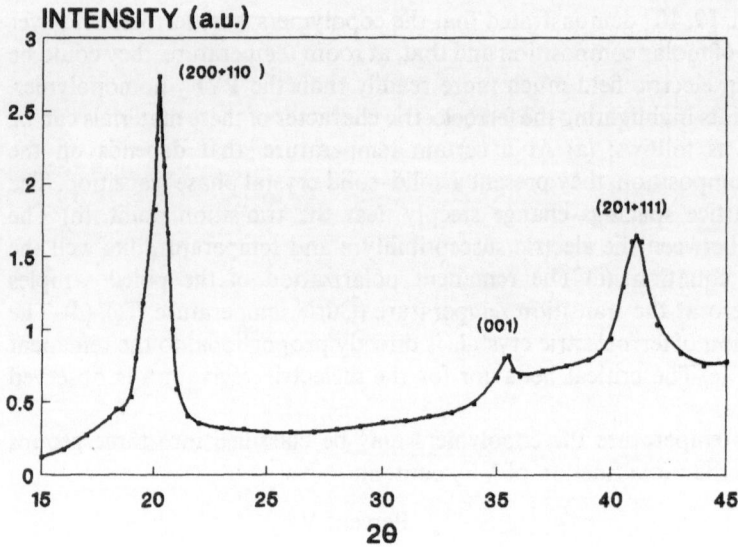

Fig. 6. Wide angle X-ray diffraction pattern for the 80/20 copolymer

Table 1 summarizes the unit cell dimensions at room temperature reported in the literature as a function of the VF_2 content. For composition of VF_2 higher than 80% the α-phase of PVF_2 is obtained. In the range of VF_2 compositions between 50 and 80% a predominant phase (orthorhombic or monoclinic) with the chains in the polar trans-conformation similar to that of the β-phase of PVF_2, giving rise to ferroelectric crystals, is observed (see Fig. 3). In the case of the 55/45 copolymer [60], the dimensions of the two coexisting unit cells – the ferroelectric and non-ferroelectric one – at room temperature are given in Table 1. Figure 7 shows the coexistence of the [110] lattice spacings corresponding to the two phases (ferroelectric and non-ferroelectric) over the whole

Table 1. Unit cell dimensions and phase assignment of the VF_2/F_3E copolymers

Sample	a	b	c	unit cell		phase	Ref.
α-PVF_2	9.63	5.02	4.62	orthorhombic		F_α	[37]
β-PVF_2	8.47	4.90	2.56	orthorhombic		F_F	[37]
91/9	9.59	4.98	4.66	orthorhombic		F_α	[37]
83/17	8.84	5.03	2.54	orthorhombic		F_α	[37]
73/27	8.94	5.16	2.56	orthorhombic		F_F	[65]
70/30	9.0	5.16	2.55	orthorhombic		F_F	[72]
65/35	8.92	5.17	2.55	monoclinic	$\beta = 93$	F_F	[73]
55/45	9.16	5.43	2.53	monoclinic	$\beta = 93$	F_{NF}	[60]
	9.12	5.25	2.55	monoclinic	$\beta = 93$	F_F	[60]
52/48	9.12	5.22	2.55	monoclinic	$\beta = 93$	F_F	[73]
37/63	9.37	5.52	2.53	monoclinic	$\beta = 93$	–	[34]
13/27	9.53	5.60	2.53	monoclinic	$\beta = 93$	–	[34]
PF_3E	9.73	5.62	2.29/ 2.50	Hexagonal		–	[34]

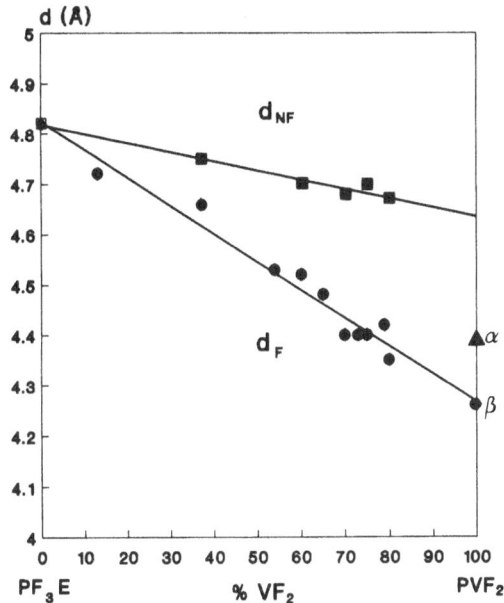

Fig. 7. Lattice spacings d_{110} of the ferroelectric, d_F, and non ferroelectric, d_{NF}, phases as a function of the VF_2 content (Data from Refs. [37, 60, 72, 79, 81])

composition range. For any copolymer composition the spacings of the polar phase d_F are smaller, and linearly decrease at a higher rate, than the d_{NF} spacings when the VF_2 content increases. This indicates that, both, the F_3E and the VF_2 units are incorporated within the crystals and that the copolymers cocrystallize isomorphically in both phases. The data superimpose for the PF_3E homopolymer giving a single non-polar spacing.

It should be noted that there is some controversy concerning the interpretation of the structure corresponding to the non-ferroelectric phase. Whereas Tashiro et al. [60, 73] suggest a disordered structure of microdomains with an all-trans conformation within the crystallites, according to Lovinger et al. [38, 58] and Lando et al. [33, 37, 56] the chain structure approximates a regular succesion of tg^+, and tg^- sequences giving a 3/1 helix. Lovinger has suggested that the planar trans-conformation would be favoured in molecular regions rich in VF_2 while those containing mostly F_3E units would favour the 3/1 helix (the 3/1 helix has a meridional peak at 2.1 Å and the copolymers have a broad maximum at 2.3 Å). According to Lovinger these two conformations would be mixed along the chain in a disordered manner so-that the overall chain structure may consist of a regular sequence of tt, tg^+, and tg^- units.

The coexisting two phases – F_F and F_{NF} – for these copolymers can be transformed into a single all-trans ferroelectric phase by using the following methods:

Uniaxial drawing in low VF_2 content samples causes not only a molecular orientation of the material, but also a transformation of the F_{NF} phase into an oriented all-trans ferroelectric phase (the reflection at 2.3 Å disappears and the doublet in the X-ray pattern at around 19° (2θ) is transformed into a single peak). However, in samples with high VF_2 content no significant changes in the structure of these copolymers are observed because these samples are already crystallized directly into the F_F phase.

Another possibility to transform the non-ferroelectric into the ferroelectric phase is by *application of an electrical field* [38]. Poling at room temperature causes a transformation similar to that obtained by drawing [38]. However, compositions with high VF_2 content which do not contain a minimum amount of disorder as that found in the melt solidified samples, show no significant improvement of molecular packing as a result of poling treatment [59, 60].

Annealing treatment above the Curie temperature influences the microstructure [63] as well as the electro-mechanical properties and switching behaviour [82] of these copolymers. After annealing, the films become brittle and the degree of crystallinity increases. Ohigashi and Koga have shown [83] that annealing above T_c results in an increase of the piezoelectric and pyroelectric properties. The annealing effects change slightly with molar composition or sample preparation. On the contrary, little changes in the structure are observed when annealing is performed at temperatures below T_c. Fernández et al. have followed the morphological changes in the lamellar structure of samples annealed in the vicinity of the melting temperature [63]. Annealing at temperatures higher than 140 °C makes the maximum in the SAXS pattern disappear. Thick lamellar structures having long periods higher than 4000 Å are proposed to explain their results.

7 Temperature-Induced Phase Changes

7.1 Calorimetric Investigations

On heating, the thermograms of VF_2/F_3E copolymers from group B (VF_2 percent ~ 12–82%) show, at least, two endothermic peaks: one at the Curie transition, T_c(up), and the other one at the melting temperature T_m (Fig. 8). The peak at T_c(up) often splits into two or three peaks that eventually can be converted into a sharp single peak by poling the samples with a strong electric field. These copolymers present a thermal hysteresis behaviour. During cooling, the exothermic peaks appearing upon solidification, T_s, and at the Curie temperature, T_c(down), are shifted towards the lower temperature side, as can be seen in Fig. 8. The multiple peak structure at the Curie transition is observed more clearly in the cooling process.

In copolymers of group B the Curie transition appears in the differential scanning calorimetric curve as a broad peak, extending over a temperature interval which narrows down with increasing VF_2 content. The transition

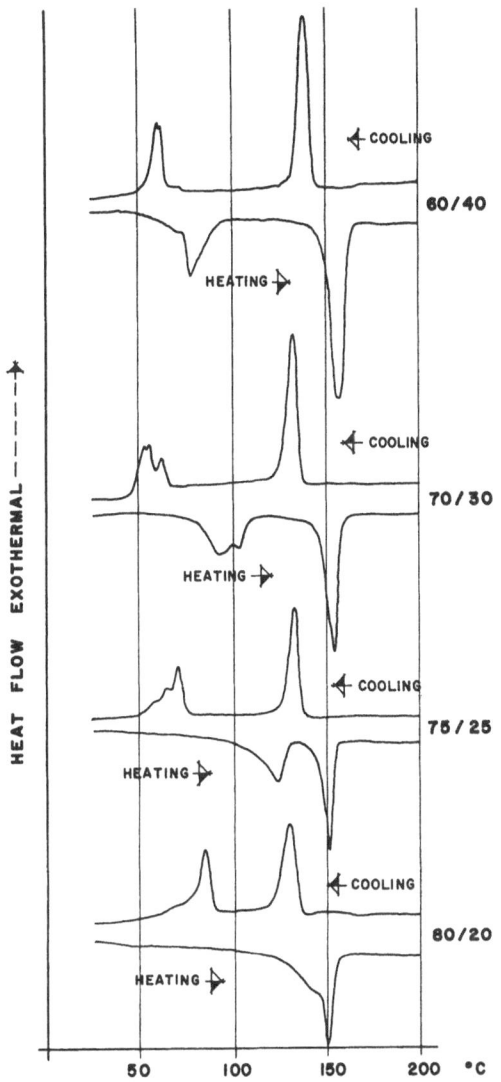

Fig. 8. DSC curves for the heating and cooling cycles of the 60/40, 70/30, 75/25 and 80/20 copolymers

extends over a wide temperature range of at least 30 °C in the 65/35 copolymer, but only of 15 °C for the 78/22 composition. The smaller transition range obtained as the number of VF_2 groups increases can be attributed to the influence of the melting peak occurring close to the Curie point.

A schematic phase diagram summarizing the three temperature regions (F_F, F_{NF} and melt) is shown in Fig. 9. For VF_2 compositions below 82%, at room temperature, one observes the predominant ferroelectric phase. With increasing temperature, the paraelectric phase appears and at higher temperatures one obtains the molten state of the paraelectric crystallites. The T_m values of the copolymers are considerable lower than those of both homopolymers and show

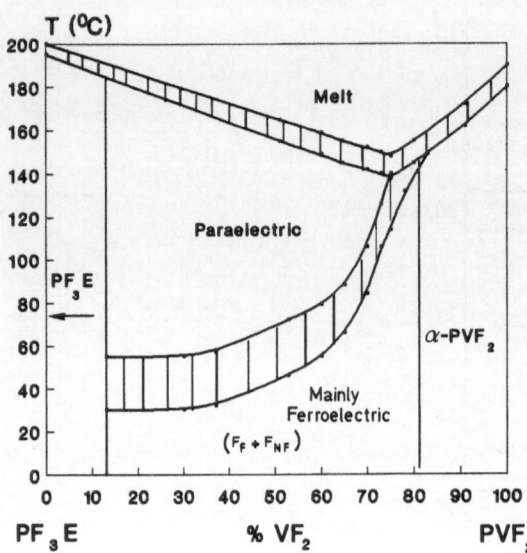

Fig. 9. Phase diagram showing the temperature versus the copolymer composition. The *dashed areas* correspond to the transition region. The frontiers among the different regions depend on sample preparation as well as on sample thermal and processing history. The fraction of ferroelectric phase, F_f, increases with increasing PVF$_2$ content

a minimum at around 145 °C in the temperature versus VF$_2$ mole fraction curve. For VF$_2$ compositions larger than 82% only the α-phase of PFV$_2$ appears, while for VF$_2$ compositions lower than 12% the PF$_3$E structure is expected. The frontiers among the different regions in the phase diagram are diffuse. In addition T_c depends on sample preparation as well as on the subsequent treatment. The dashed area corresponds to the transition region between neighbouring phases. The value of T_c conspicuously depends on VF$_2$ content. T_c(up) increases first slowly with increasing VF$_2$ content. Thereafter, it increases more rapidly when the VF$_2$ concentration is higher than 50%. Finally, T_c(up) coincides with the T_m value for VF$_2$ compositions equal or larger to 0.82. The temperature difference $\Delta T_c = T_c$(up) $- T_c$(down) (not shown in Fig. 9) increases with VF$_2$ content, from values of 10 °C for 37/63 to more than 50 °C [83] in the 80/20 copolymer.

Drawing [39], annealing [63], poling [59], hydrostatic pressure [64], electron irradiation [65–67], solution history [68] and tensile stress [69], among others, are physical treatments which can affect the Curie temperature. For a given copolymer composition, the Curie temperature can vary over a 20 °C temperature range owing to the influence of different thermal and processing histories. Most specially, the Curie temperature and the enthalpy of the transition are influenced by the crystallization temperature as shown by Tanaka et al. [70, 71]. This variability is attributed, in part, to the changes in local concentration of F$_3$E comonomer units within the crystalline lattice.

The melting temperature and melting enthalpy are, however, unchanged under poling and are only slightly affected by crystallization conditions since the influence of the preparation conditions is lost during the Curie transition. It is noteworthy that the enthalpy at the Curie transition Δh(up) increases with the

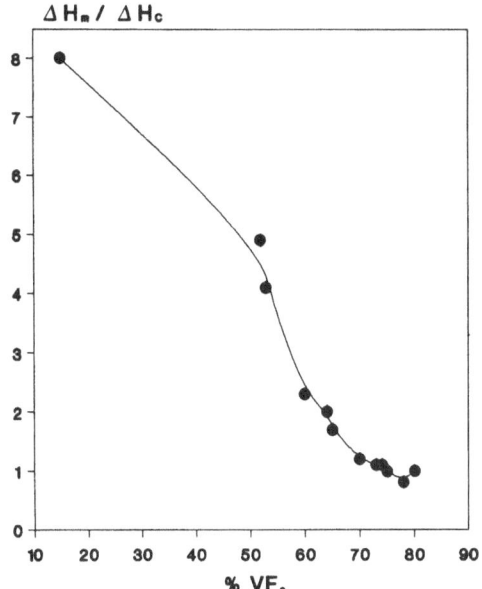

Fig. 10. Dependence of the ratio (melting enthalpy)/(Curie enthalpy), $\Delta h_m/\Delta h_c$, with VF_2 mole fractions in %

increase of VF_2 content and reaches values which are similar to those of the melt enthalpy Δh_m for VF_2 mole fractions between 70 to 80%. Figure 10 shows the decrease of the ratio $\Delta h_m/\Delta h_c$ (up) with increasing % VF_2 yielding, indeed, a value of $\Delta h_m/\Delta h_c$ (up) ≈ 1 for mole fractions of 70–80% VF_2. From this, it is apparent that the higher the Curie point (higher VF_2 content) the lower is the $\Delta h_m/\Delta h_c$ (up) ratio. Parallel behavior to $\Delta h_m/\Delta h_c$ (up) is observed for $\Delta h_s/\Delta h_c$ (down), with values only slightly larger in the latter case.

7.2 Influence of Pressure

The pressure dependence of T_c for the 65/35 composition was investigated by Koizumi et al. [64]. These authors found that the pressure dependence of T_c follows the equation:

$$T_c(K) = a + b \cdot p(MPa) \tag{3}$$

where $a = 359$ and $b = 0.383$ during heating, and $a = 333$, $b = 0.405$ during cooling. For the copolymer 52/48 $a = 340$ and $b = 0.383$ and little difference was found between heating and cooling.

Neutron diffraction studies under pressure [84] on the 70/30 composition have revealed that transitions in this copolymer are displaced towards higher temperature with increasing pressure, as can be seen in the phase diagram of Fig. 11. In addition, it is worth noting the non-linear increase of the Curie temperature with pressure. By considering the Clausius-Clapeyron relation: $dT_c/dP = T_c \Delta V_c/\Delta h_c$, this effect can be related to a decrease in the volume

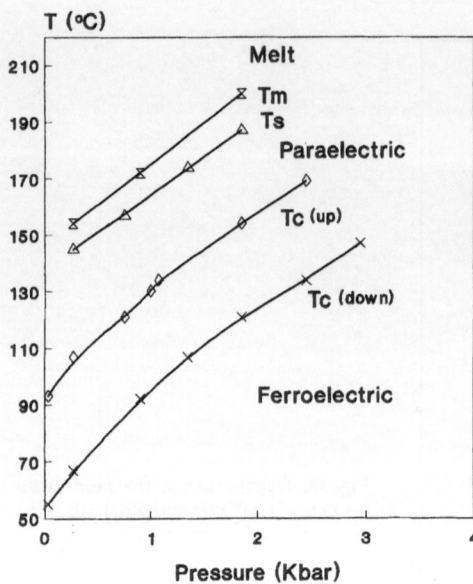

Fig. 11. Pressure dependence of T_c(up), T_c(down), T_m and T_s for the 70/30 copolymer (Figure from Ref. [84])

change ΔV_c or to an increase of the enthalpy change Δh_c under increasing pressure. The determination of the lattice parameters indicates that the volume change with pressure is almost constant. This implies that the enthalpy change would be about twice as large at 2 Kbar than at atmospheric pressure.

7.3 Changes in Structure Upon Heating and Cooling

Detailed structural studies of copolymers above and below the transition temperature using X-ray diffraction, have been performed by Lovinger et al. [38, 39], Tashiro et al. [34, 65, 73, 74], Legrand et al. [72, 75], Yamada et al. [9, 10] and in our laboratory [62, 76], in unoriented and oriented melt-solidified samples. Figure 12 illustrates as an example, the change of the WAXD pattern with temperature in real time for the copolymer 60/40 in the angular range $16° \leq 2\theta \leq 21°$. The sample was first heated at a constant rate of 10 °C/min from room temperature (25 °C) up to 170 °C. Then it was kept at the same temperature for 5 min and finally it was cooled down at the same heating rate to room temperature.

The diffraction pattern, at room temperature, shows a superposition of a broad peak associated to the ferroelectric phase centered at $2\theta = 18.97°$ and a shoulder at $2\theta = 18.36°$ corresponding to the non-ferroelectric phase. As the temperature is increased, the intensity of the fainter peak increases and that of the ferroelectric maximum decreases concurrently. At the Curie temperature, the peak characteristic for the ferroelectric structure disappears and only the reflection corresponding to the paraelectric structure is present. On further

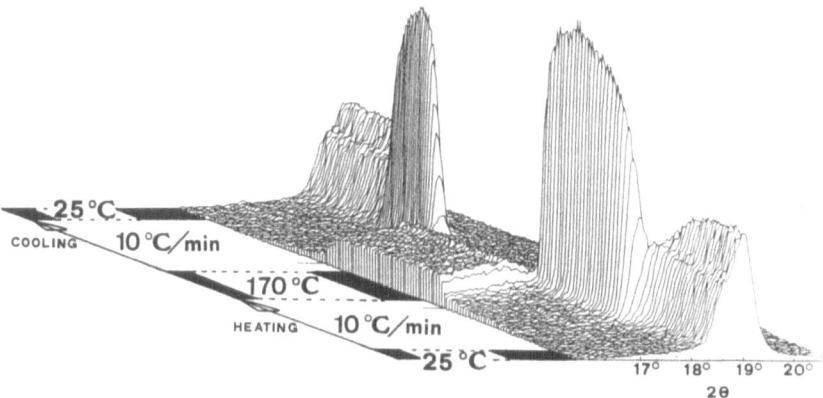

Fig. 12. Three dimensional plot of the WAXD patterns for the 60/40 copolymer as a function of temperature during heating followed by cooling

heating the intensity of the reflection arising from the paraelectric structure rapidly increases until the vicinity of the melting point. Beyond T_m the appearance of a broad halo centered at 5.3 Å indicates the onset of melting. When the sample is cooled down, the opposite process is observed. However, owing to the thermal hysteresis the reappearance of the ferroelectric peak is shifted towards lower temperature in agreement with the DSC results (Fig. 8). At room temperature the diffraction pattern resembles that of the starting material.

The typical behavior of the lattice spacings for various copolymers during crystallization from the melt, as a function of temperature is shown in Fig. 13. At temperatures above the melting temperature the d-spacing is derived from the angular position of the amorphous halo and remains constant at about 5.2–5.3 Å depending on composition. Below the temperature of solidification the reflection corresponding to the lattice spacing of the paraelectric phase, $d_p \approx 5$ Å, appears. In the temperature region between the temperature of solidification and T_c (down), d_p decreases linearly with T. At the Curie temperature not only a new spacing, d_F, corresponding to the ferroelectric phase appears but also a small fraction of the initial paraelectric crystals are gradually transformed into the new non-ferroelectric phase with a spacing d_{NF}. Finally, in the low temperature range, below T_c (down), the spacings d_F and d_{NF} corresponding to the two coexisting phases are observed. The stepwise behavior for the d-spacing is most prominent for the 60/40 copolymer. For the other copolymers the stepwise-behavior of the d-spacing at the Curie point is less conspicuous when the VF_2 concentration increases. The observed changes in lattice spacing during the ferroelectric-paraelectric transition are correlated with conformational changes through introduction of g^+ and g^- bonds leading to a random sequence of tg^+ and tg^- groups in the paraelectric phase which is similar to the disordered 3/1 helical conformation of the PF_3E [38]. In the paraelectric phase, the chains assume this disordered conformation and are packed on a pseudo-hexagonal lattice.

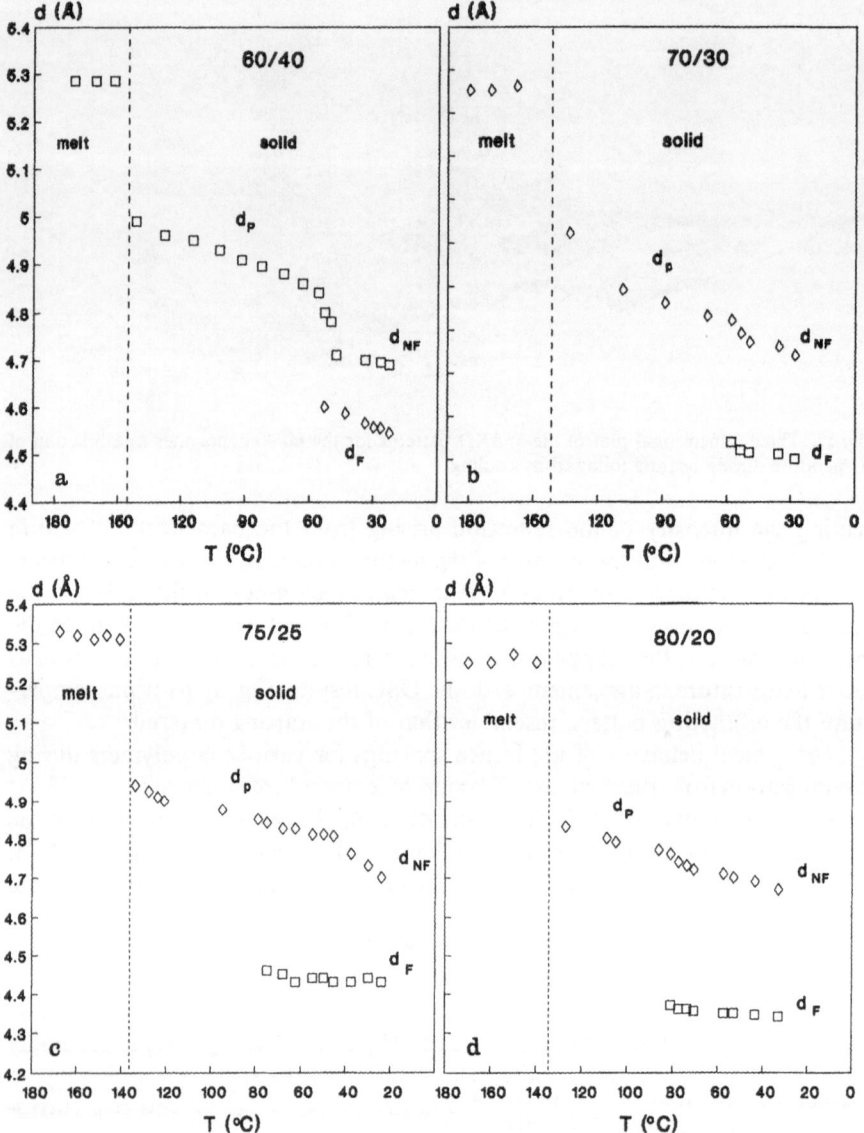

Fig. 13a–c. Changes in lattice spacings for the ferroelectric, d_F, non ferroelectric, d_{NF}, and paraelectric, d_p, phases as a function of temperature for the **a.** 60/40; **b,** 70/30; **c.** 75/25 and **d.** 80/20 copolymers during cooling from the melt. On the left side of the figures the d-spacing of the amorphous halo from the melt is shown

The ferroelectric all-trans diffraction maximum is always significantly broader than its paraelectric counterpart. Figure 14a shows as an example the variation of the integral width of the WAXD peak with temperature through the Curie transition for the 60/40 copolymer. In the high temperature range, δβ

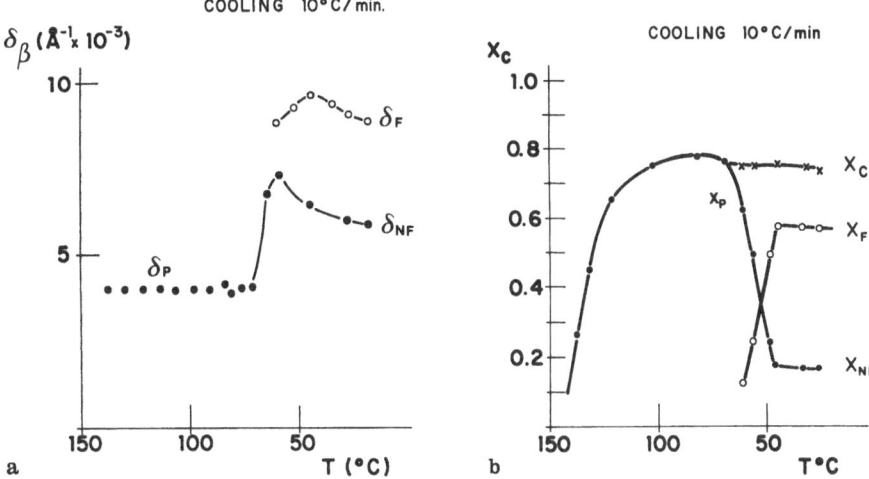

Fig. 14a, b. Variation of **a.** integral width and **b.** crystallinity, x_c (including the ferroelectric, x_c^F, non ferroelectric, x_c^{NF}, and paraelectric x_c^P, crystalline fractions) as a function of decreasing temperature

remains nearly constant. Just before T_c (down) is reached, $\delta\beta$ rapidly increases showing a maximum value at T = Tc. At this point an even broader diffraction peak corresponding to the ferroelectric phase emerges. In the low temperature range, both, $\delta\beta_{NF}$ and $\delta\beta_F$ slightly decrease with further decreasing the temperature.

Figure 14b presents, as an example, the variation of crystallinity x_c and the relative fraction contribution of the different phases during cooling with con-stant cooling rate. Below T = 150 °C, it is seen that $x_c = x_c^P$ (crystallinity of the paraelectric phase) rapidly increases with decreasing temperature and levels off at· 100 °C. At this temperature, crystallization is completed. Most interesting is the fact, that in the 60–55 °C range (Curie transition) the volume fraction of the emerging ferroelectric (x_c^F) crystals linearly increases, while a concurrent de-crease of the volume fraction of the paraelectric phase (x_c^P) which gradually changes into the non ferroelectric phase in this interval, is observed. Below T = 50 °C, the ferroelectric phase predominates ($x_c^F \approx 60\%$, $x_c^{NF} \approx 18\%$), both fractions, x_c^F and x_c^{NF} remaining constant.

7.4 Changes in the Long Period

The Curie transition can also be detected by following the small angle X-ray scattering (SAXS) maximum [77, 78]. The samples present a broad scattering maximum at room temperature. The position and the intensity of the maximum does not change substantially until the Curie temperature is reached. At T_c the intensity decreases, and the position of the scattering maximum is shifted towards smaller angles during the transition. Thereafter the scattering intensity increases with temperature up to the vicinity of the melting point were it finally drops to zero.

Figure 15 illustrates the variation of L with temperature for the heating and cooling process in the copolymer 60/40 [79]. The initial long period L $\sim$ 300 Å remains nearly constant within the error of experiment in the low temperature region. At the Curie temperature, L shows a sudden discrete increase up to 325 Å, which corresponds to a thickening of the lamellae. Above T_c, L remains again constant in the high temperature region. Finally, a very large L-increase at about T = 150 °C is observed. On cooling (at a rate of 10 °C/min) essentially a symmetrical behavior is observed for this copolymer. A maximum appears in the scattering pattern and a decrease of L with T is observed: just below T_s one can follow the first steps of crystal nucleation showing a distinct decrease of L from 410 Å to 325 Å for a change of 140 to 130 °C. In the high temperature range (130–60 °C) the value of L-325 Å practically remains constant. At T = T_c, L shows a small decrease from 325 to 310 Å confirming the stepwise behavior of L near T_c. Finally the long period shows a gradual increase with decreasing temperature up to L = 325 Å.

The above-mentioned behavior, as well as the average long spacing, seems to depend, on both the thermal treatment and the polymer composition. For instance, for copolymers with a higher VF_2 content, one observes the same L-increase on heating at T_c. However, when cooling from the melt, the scattering pattern does not show any diffraction peak. Annealing up to temperatures close to T_m produces a SAXS intensity curve similar to that observed when cooling the melt with VF_2 composition higher than 60% at 10 °C/min [78], i.e. the long spacing disappears (until 2000 Å) suggesting the onset of large morphological changes in the microstructure of the copolymer. One possibility is that the scattering curve would be the tail of a diffraction maximum which

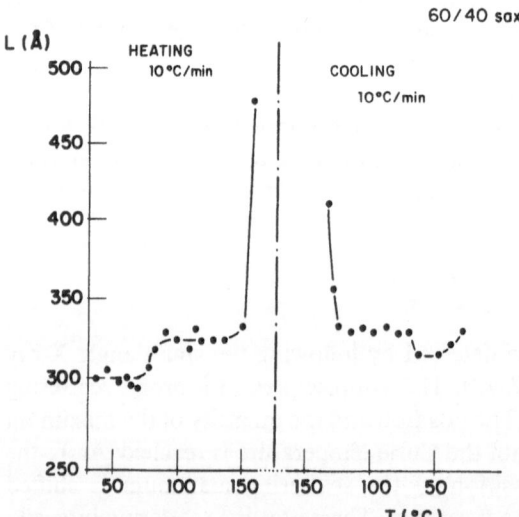

Fig. 15. Long period versus temperature as derived from the SAXS maximum using Bragg's law for the 60/40 copolymer

cannot be detected because of lack of experimental resolution. A second possibility is that after a fast cooling rate or annealing near to T_m a "frozen in" nematic liquid crystalline structure would be present. Legrand [80] has also pointed out that after annealing through the Curie temperature the regular stacking of lamellae disappears and is replaced by another superstructure which

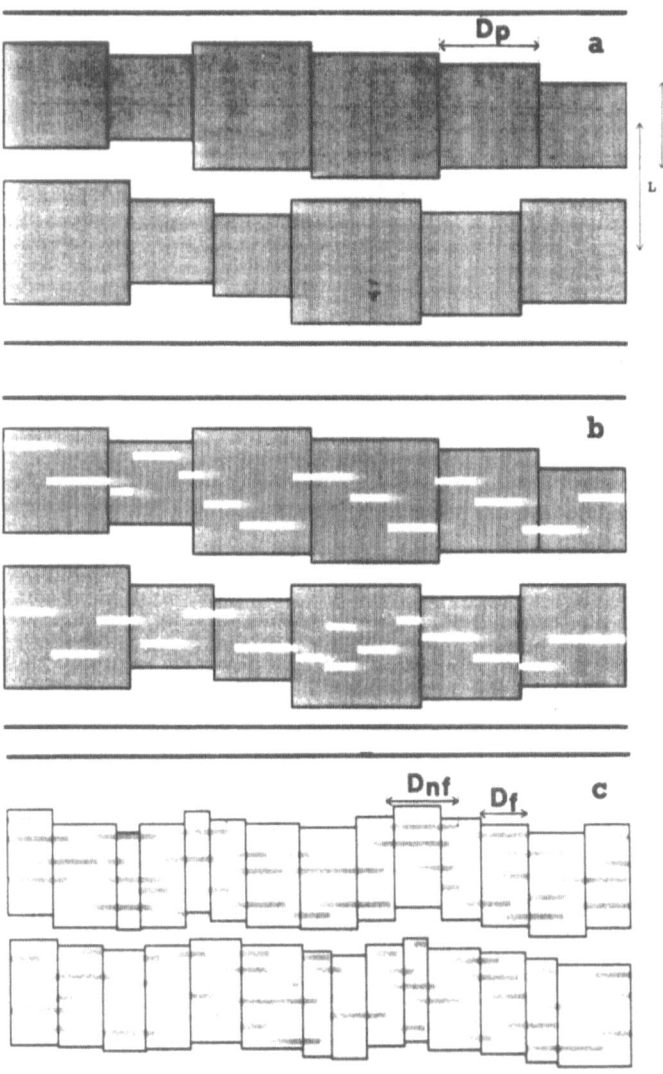

Fig. 16a–c. Schematic model of the lamellar structure of the copolymer in the, **a.** high temperature range (paraelectric phase) **b.** Curie transition region and **c.** low temperature region L and l denote respectively the long period and the average crystal thickness comprising a mixture of non ferroelectric and ferroelectric domains

accommodates better the effects of the irregularly arranged anisotropic expansion of the crystallites through the Curie transition.

7.5 Structural Model

The appearance of two diffraction peaks with different integral breadths in the wide angle diffraction pattern, at T_c during cooling (Fig. 13) has been shown to indicate that the transition involves the breakdown of paraelectric crystal blocks into smaller ferroelectric domains with lateral dimensions of about 110 Å [79]. Direct evidence for the fragmentation of the paraelectric crystallites into smaller ferroelectric ones during cooling has been given earlier by dark field electron microscopy by Lovinger [65]. A model based on SAXS and WAXD data to explain the structural changes observed during cooling the 60/40 copolymer from the melt has been recently proposed [79, 81]. A sketch of the model of the transition through T_c is shown in Fig. 16. In the high temperature region the stacks of lamellar crystals yielding a long period $L \approx 325$ Å are essentially composed of paraelectric blocks of about 250 Å in lateral dimensions. At the Curie temperature, the paraelectric crystals are gradually transformed into smaller ferroelectric domains which become statistically mixed with non-ferroelectric remaining zones. In the low temperature regions most of the material is transformed into the ferroelectric phase although a small fraction of material emerges as a non-polar structure. According to this model the X-ray long period below T_c comprises the alternation of crystal blocks, consisting of a random coexistence of ferroelectric and non-ferroelectric domains, separated by amorphous regions.

8 Mechanical Properties

8.1 Microhardness Relating to Structural Parameters

Microhardness (H) has proved to be a powerful method for the investigation of morphological and microstructural changes in semicrystalline polymers [85, 86]. The use of H has also proved effective for detecting polymorphic changes in crystalline polymers [88] and phase transformations in amorphous solids including polymers [89, 90], in particular the glass transition behavior. A general expression for the microhardness of a semi-crystalline polymer at a given absolute temperature T above glass transition, having a mean crystal thickness l and a crystallinity w_c, has been given as [87]:

$$H = w_c H_o e^{-\beta(T - T_o)}/(1 + b/l) \tag{4}$$

where β is the thermal softening coefficient, H_o is the hardness for crystals with infinite thickness at the reference temperature T_o and b is a parameter which is connected to the surface free energy of the crystals and is inversely related to the energy required for crystal destruction [91, 92].

We have seen, that the crystallinity for these copolymers, w_c, remains nearly constant as a function of temperature (see Sect. 7.3). In addition, the value of the crystal thickness, l, is also constant below and above T_c. Since H_o and b are constants, one may assume that, for these copolymers, the quantity $w_c H_o/(1 + b/l) = $ const for each composition. Hence, according to Eq. 4, for a given crystalline phase, one can essentially expect an exponential variation of H as a function of T.

8.2 Dependence of Microhardness on Temperature

Figure 17 (top) shows the obtained H-variation as a function of increasing temperature for the three investigated copolymers with 60/40, 70/30 and 80/20 mole fractions. It is seen that log H decreases linearly with increasing temperature. The onset of the Curie transition can be clearly identified with a bend in the $\log(H) - T$ plot. On cooling (Fig. 17 bottom) H increases back with decreasing temperature, and the point at which the Curie transition is ended can be again identified with the change in slope of the plot.

For simplicity, it is convenient to describe the changes in log(H) in the light of the results obtained with the 60/40 and 70/30 copolymers during heating. In both cases one may distinguish 3–4 different regions in the $\log(H) - T$ plot, suggesting the presence of different structural mechanisms. The lowest and highest temperature regions correspond to the thermal expansion of the ferroelectric and paraelectric crystalline phases respectively (Fig. 17 open symbols) and are characterized by two thermal softening coefficients β_F and β_P. Within the Curie transition region (solid symbols) the presence of two further slopes in the $\log(H) - T$ plot can be inferred. These two regions might be associated with the doublets appearing in the calorimetric scans observed in the Curie transition region. To explain the first slope (90–100 °C interval for the 70% VF_2 copolymer) one might suggest a transition mechanism from the ferroelectric to the non-ferroelectric phase described by Tashiro and Kobayashi [60]. For the second slope (100–105 °C region for the 70% VF_2 copolymer) one might think of a transition from the non-ferroelectric phase to the final high temperature paraelectric phase. In case of the 80/20 copolymer, in addition to the initial H-decrease corresponding to the thermal expansion of the ferroelectric phase, one observes the onset of the Curie transition at about 115 °C which is, then, masked by the melting of the first crystals.

On cooling, the opposite behavior of the microhardness, increase with decreasing temperature, is observed. However, the Curie transition temperatures are shifted towards lower temperatures.

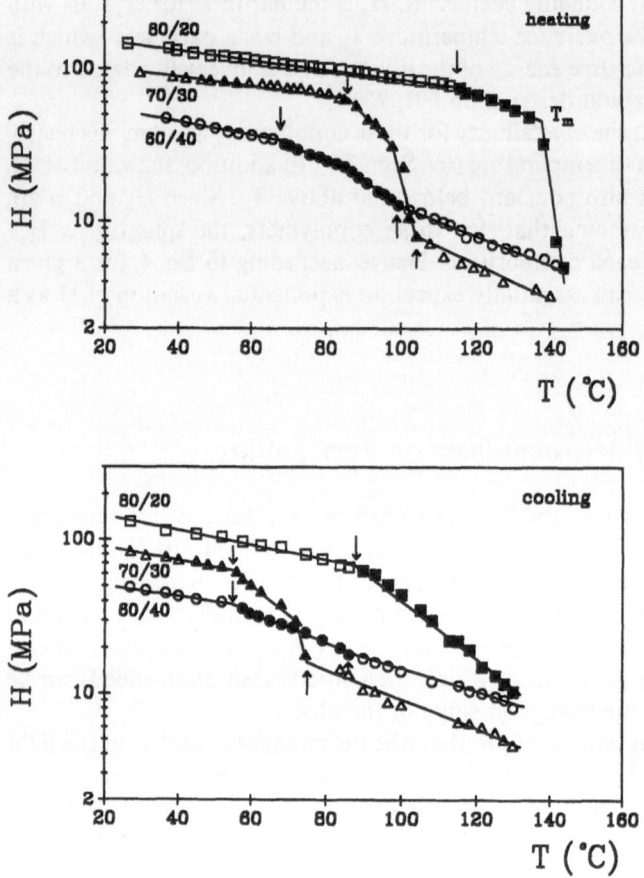

Fig. 17. Temperature dependence of the microhardness during heating and cooling of the different copolymers

8.3 Influence of Copolymer Composition

Most interesting is the fact that H at room temperature (Fig. 17) appreciably increases with VF_2 content. Since the fraction of ferroelectric crystals is nearly constant ($x_c \sim 0.5$–0.6) for the various compositions [81] the hardness increase at low temperature can be explained on the basis of the structural characteristics of the ferroelectric phases of β-PVF_2 and its copolymers. In β-PVF_2 the 4.91 Å spacing is determined by the closest possible chain packing in the b-direction. Since this is also the dipolar direction, the chains are held together by strong attractive forces so that any intramolecular rotation will be sterically hindered. In the copolymers, on the other hand, the separation between chains along b increases with increasing CFH units randomly located along the chain axis. In addition, with increasing F_3E sequences the overall dipole moment within the

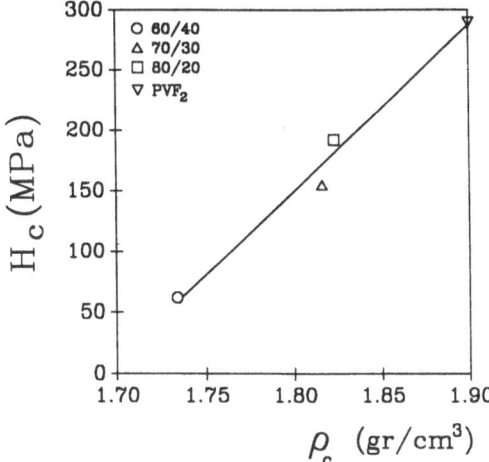

Fig. 18. Variation of crystal hardness, measured at room temperature, as a function of unit cell density for the various VF_2/F_3E compositions

ferroelectric crystal will diminish. As a result, the crystal hardness (critical stress to deform the crystals irreversibly) decreases with increasing CFH units and shows a maximum value for chain packing. Figure 18 illustrates the linear increase of the crystal hardness with increasing unit cell density of the co-polymers in the ferroelectric phase, including the highest H_c-value found for the PVF_2 homopolymer.

8.4 Dependence of Microhardness Upon Crystal Density

Figure 19 illustrates the dependence of the crystal hardness upon crystal density for various compositions. This plot permits to distinguish between the two

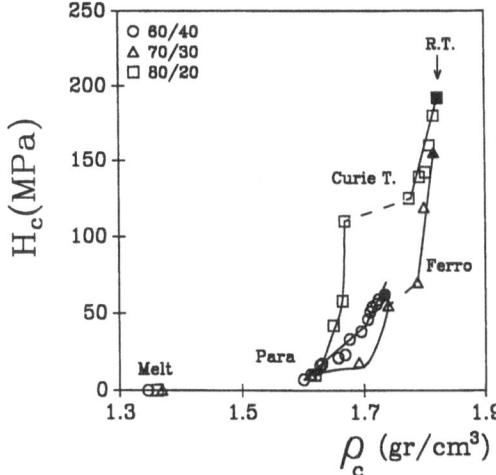

Fig. 19. Dependence of crystal hardness upon unit cell density through the Curie transition

distinct contributions to H_c: (a) a gradual lattice contraction and (b) a drastic change from the paraelectric to the ferroelectric phases during cooling. On cooling a lattice contraction takes place (p_c increases continuously) and H_c for each copolymer gradually increases. At the Curie transition the crystal density, p_c, for the 70/30 and 80/20 compositions suffer a discontinuous increase contributing to a slight H_c increase. For the 60/40 composition one observes, at the Curie transition, just an inflexion point in the H_c vs p_c plot. In the ferroelectric phase the lattice contracts further on cooling for the three copolymers, giving rise to a very steep crystal hardness increase with increasing crystal density.

8.5 Hysteresis Behavior

We wish finally to discuss the H behavior of the copolymers concerning the question as whether H varies reversibly with temperature. Figure 20 illustrates, as an example, the evolution of a H–T cycle during heating and cooling for the 70/30 copolymer using the data of Fig. 17. The fact that on cooling the Curie temperature range is shifted towards lower temperatures leads to a corresponding shift of the paraelectric phase lower boundary from 105 °C down to 75 °C. Within the Curie region, H increases further up to the inflexion point near T = 55 °C where the ferroelectric phase is restored. However, the slightly lower H values found after solidification could suggest the persistence of certain conformational tg^+, tg^- sequences anchored within the ferroelectric crystals which reduced dipole alignment within the chains and which are finally removed at room temperature. A similar hysteresis behavior is suggested for the 60/40 copolymer, though the H-changes are very small in this case. In the case of the 80/20 copolymer, the hysteresis H-curve cannot be fully obtained due to the superposition of the Curie transition and the melting of crystals beyond

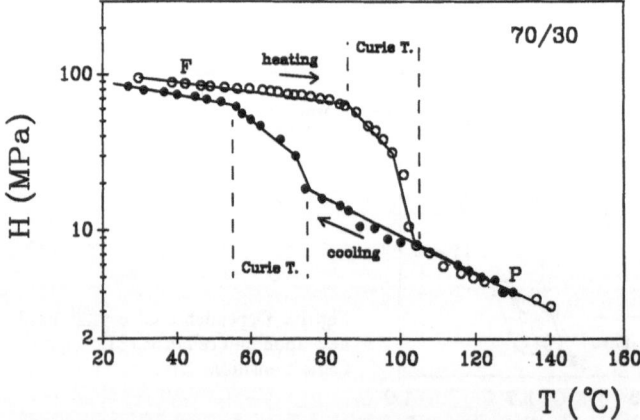

Fig. 20. Thermal hysteresis behavior of log(H) vs T for the 70/30 VF$_2$/F$_3$E copolymer

T = 140 °C. Here, during solidification, the H increase from 140 °C down to about 100 °C is the result of a double contribution of: (a) the crystallization of the fraction of molten crystals and (b) the thermal contraction of the nonpolar phase crystals. The hysteresis behavior is also found in other mechanical properties (dynamic modulus) derived from micromechanical spectroscopy [66, 67], where it is shown that the hysteresis cycle shifts to lower temperatures if the samples are irradiated with electrons. It has also been pointed out that the samples remain in the paraelectric phase, when cooling, if the irradiation dose is larger than 100 Mrad.

In conclusion, it appears that microhardness yields information about paraelectric to ferroelectric phase changes in VF_2/F_3E copolymers which can be discussed in the light of the changes in the lattice spacings of the different phases and in variations of the crystallinity value.

9 Electrical Properties

9.1 Polarization

A common technique for obtaining macroscopic polar films involves poling by electrical fields. In this case the polarization induced in the sample remains after the removal of the electric field. Figure 21 shows the variation of the polarization of a film of the 55/45 copolymer as a function of an applied electric field [8]. The value of the remanent polarization, P_r, of samples measured in our

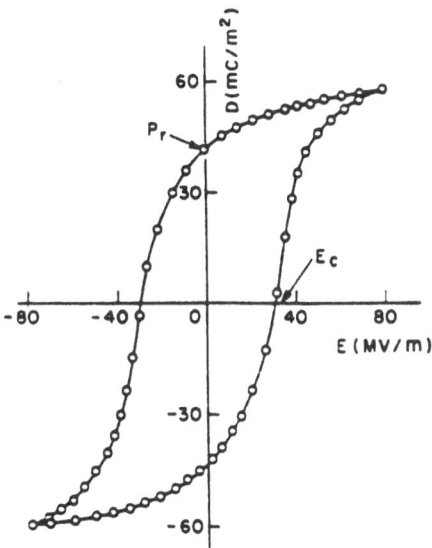

Fig. 21. Electric displacement, D, versus electric field, E, hysteresis loop of the 55/45 copolymer at 20 °C (Figure from Ref. [8])

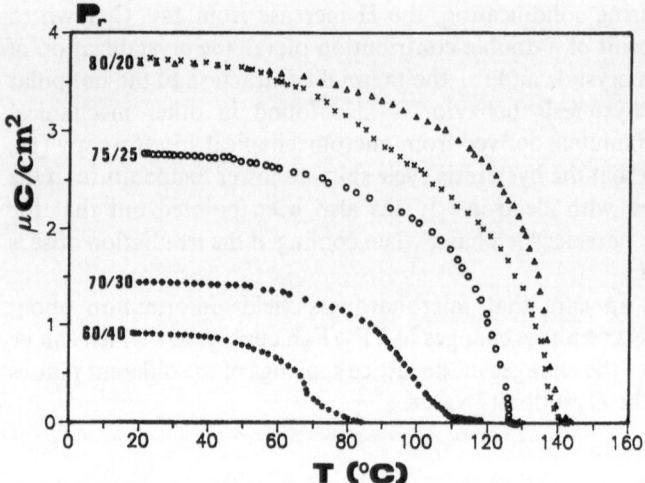

Fig. 22. Remanent polarization as a function of temperature for various copolymers. The two curves for the 80/20 correspond to two different crystallinities. The (x) symbols correspond to samples with lower crystallinity

laboratory depends on temperature as illustrated in Fig. 22. Here an electrical field of 500 kV/cm has been applied during 15 minutes at room temperature. The value of P_r decreases as the temperature increases and disappears at a certain temperature. The temperature at which P_r become zero has been associated with the occurrence of the ferroelectric to paraelectric phase transition (see Fig. 16) [17].

Most important is, however, the fact that P_r strongly depends on VF_2 content. This is because the copolymers adopt the all-trans highly polar conformation (see Sect. 3) and with increasing number of VF_2 units the resulting dipole moment within each crystal increases. In fact, it has been shown that the electric polarization in these copolymers increases with the fraction of ferro-electric crystals in the material.

Another technique used for obtaining macroscopically polar films involves mechanical extension of the material. Uniaxial plastic deformation induces a destruction of the original spherulitic structure into an array of crystallites in which the molecules are oriented in the deformation direction. In case of PVF_2 when such deformation takes place below 90 °C the original tg$^+$ tg$^-$ chains are forced into their most extended possible conformation which is all-trans [32].

9.2 Dielectric Behavior

Dielectric spectroscopy has been shown to be of great interest in dealing with transitions involving reorientation of permanent dipoles [93]. By monitoring the temperature and frequency dependence of the complex dielectric permittivity, $\varepsilon^* = \varepsilon' - i\varepsilon''$, substantial information can be gained relating the molecular dynamics of the specimen under study [94]. This procedure has been successfully applied to investigating polymeric materials [93–95].

In particular, VF_2/F_3E copolymers have also been the subject of extensive research [6, 17, 96]. As an example to illustrate the dielectric behavior of these copolymers, the temperature dependence of the real and the imaginary part of the complex permittivity at two different frequencies (1 and 100 kHz) are shown in Figs. 23a and 23b respectively. The measurements correspond to the 60/40 copolymer. The data have been collected by using a sandwich geometry with gold evaporated electrodes [95]. Frequencies of 10^3 and 10^6 Hz have been used by employing a 4192 A HP Impedance Analyzer. From inspection of Fig. 23b

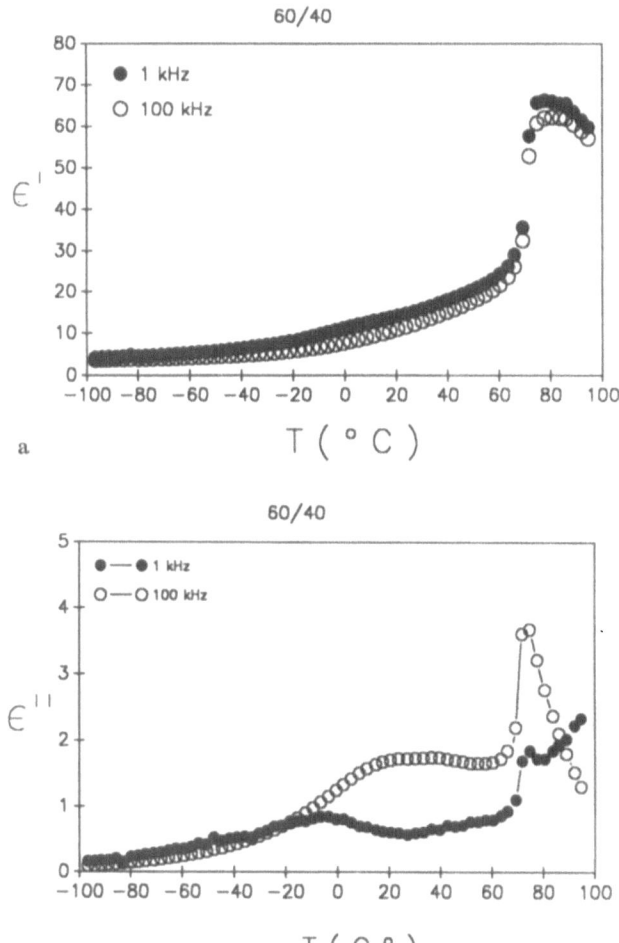

a

b

Fig. 23a, b. Isochronal plot of **a.** the real part, ε', and of **b.** the imaginary part, ε'' of the complex permittivity at two frequencies for the 60/40 copolymer

two main relaxation processes are clearly detectable in covering the temperature range. In Fig. 23a, the two processes can also be identified as an increment of ε' with T, although the low temperature process is less conspicuous. These transitions are commonly designed as "T_t"[1] and "β" in order from higher to lower temperature. The β-relaxation shifts towards higher temperatures as the frequency is increased. The T_t-relaxation, in contrast, exhibits no temperature dependence in the frequency range investigated. These effects can be better visualized on the three dimensional plots shown in Figs. 24a and 24b. Here the dependence of ε' and ε'' versus temperature and frequency is represented. At

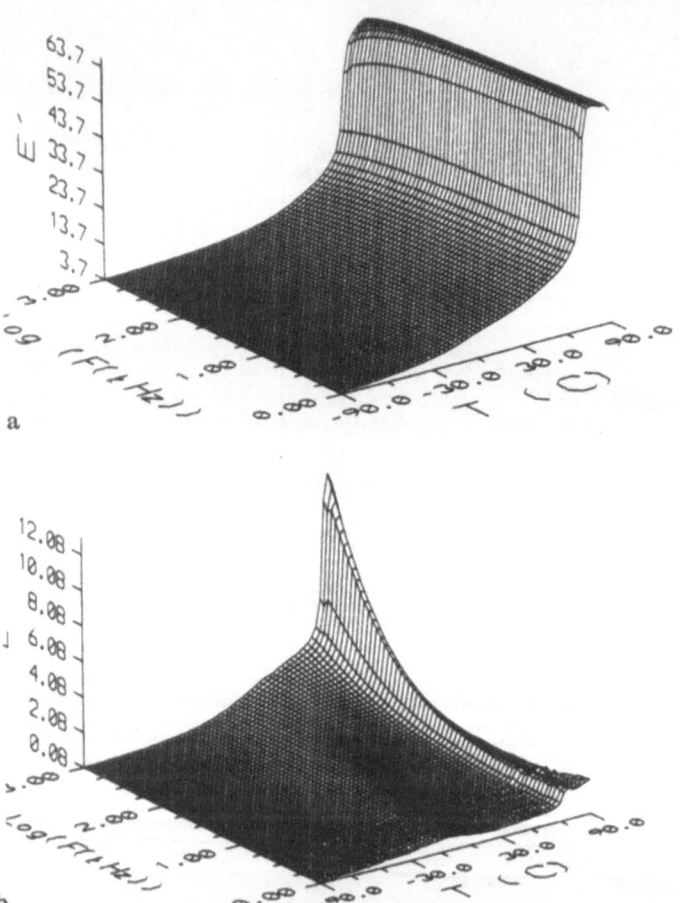

Fig. 24a. ε' and **b.** ε'' vs temperature and frequency for the 60/40 cópolymer

---

[1] We denote the first transition "T_t" instead of α-transition to distinguish it from the α-transition derived from mechanical dynamical experiments.

lower temperatures $(T < -100\,°C)$ an additional relaxation process "γ" has been reported [96].

Thorough analysis of both T_t and β relaxations has been reported for concentrations of VF_2 larger than 55% mol [96]. The origin of the two low temperature relaxations, β and, γ seems to be similar to those appearing in PVF_2 [6]. In particular, the "β" process has been attributed to the segmental motion in the amorphous phase of the semicrystalline polymer and, consequently, it has been correlated to the glass transition T_g. However, subsequent experiments have shown that the crystal-amorphous interface also may play an important role in the β-process in PVF_2 [97]. The observed high temperature transition, T_t-process, has been attributed to the ferro-paraelectric transition [17]. As we have seen, near the Curie transition, the onset of rotational motions in the chain stems within the crystals provokes a considerable number of tg^+, tg^- conformational isomers which facilitates the change from the ferroelectric β-phase to the paraelectric phase [83, 98].

As an example of the Curie transition for other VF_2/F_3E compositions, measurements of $ε'$ and $ε''$ as a function of temperature and $f = 1$ kHz are shown in Figs. 25a and 25b. The complex permittivity function of polymeric materials has been shown to follow the Havriliak-Negami phenomenological equation [99]:

$$ε^* - ε_∞ = Δε/(1 + (iwτ_0)^a)^b \tag{5}$$

with $Δε = ε_∞ - ε_0$, being $ε_∞$ and $ε_0$ the unrelaxed and relaxed dielectric constants, $w = f·2π$ the angular frequency and $τ_0$ the central relaxation time. This equation generalizes the Cole-Cole and the Davison-Cole equations including symmetric and asymmetric broadening accounted for by the a and b parameters respectively [95]. Values of $b = 1$ (symmetric diagram) have been reported to render good fits for the VF_2/F_3E copolymers with VF_2 content larger than 55% [96].

In Fig. 26 some selected data from Figs. 23 and 24 (60/40 copolymer) are represented in a Cole-Cole diagram. The continuous lines represent the corresponding theoretical fits according to Eq 1. The corresponding parameters, found using a least square procedure, are represented in Table 2. As it is shown, in this case, too, good agreement between theory and experiment can be achieved with $b = 1$. From inspection of the fitting parameters, a decrease of the central relaxation time with increasing temperature is observed. Simultaneously a slight increase of the symmetric broadening parameter "a" with increasing temperature is also found. The measurements performed in the case of the 55/45 copolymer [96] point towards the existence of a monodisperse relaxation process $(a = 1)$ when the critical transition temperature is approached. This effect is accompanied by a slowing-down of the central relaxation time at the temperature region where the ferro-paraelectric transition takes place. Although several theoretical approaches have been proposed to explain the dielectric relaxation associated to the ferro-paraelectric transition until now only qualitative agreement has been found [8, 83].

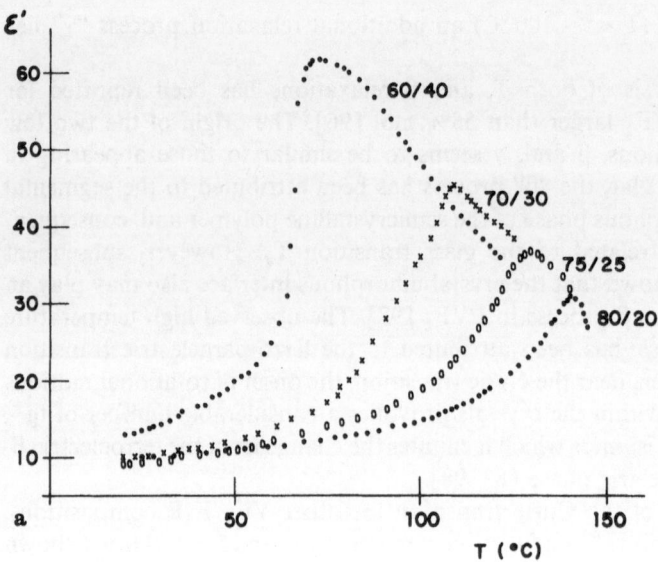

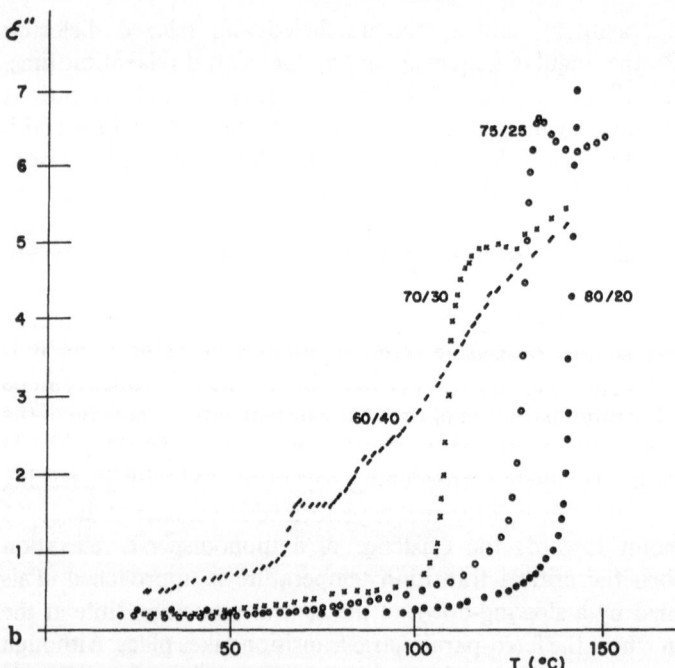

Fig. 25a, b. Temperature dependence of **a.** the real part, ε′ and of **b.** the imaginary part, ε″ of the complex permittivity for copolymers with different compositions

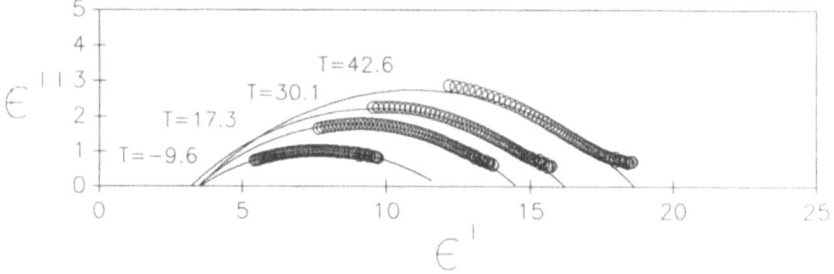

Fig. 26. Cole-Cole plots at some selected temperatures T (in °C) for the 60/40 copolymer

Table 2. Fitting parameters of Havriliak-Negami model

$\Delta\varepsilon$	b	a	ε_∞	$\tau_0(s)$	$T(°C)$
8.4	1	0.3058	11.91	$6.4*10^{-6}$	− 9.6
11.1	1	0.3968	14.53	$4.7*10^{-7}$	17.3
13.02	1	0.4186	16.22	$1.6*10^{-7}$	30.1
15.45	1	0.4430	18.65	$8.0*10^{-8}$	42.6

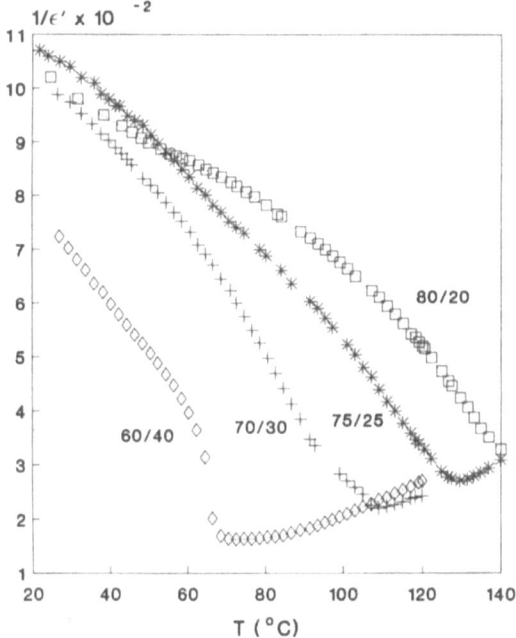

Fig. 27. $1/\varepsilon'$ as a function of temperature for copolymers with different compositions

It is interesting to discuss, next, the available ε' data for the various compositions in the light of the classical theory of ferroelectricity. According to this theory [100] for $T \geq T_c$ the temperature dependence of ε' can be written as

$$\varepsilon' = 1 + A/(T - T_o) \qquad (6)$$

in CGS units, where $T_o < T_c$ or $T_o = T_c$ depending on the order of the transition. In the proximity of T_o, $1/\varepsilon'$ follows a Curie-Weiss law

$$1/\varepsilon' \approx (T - T_o)/A \qquad (7)$$

In Fig. 27 experimental $1/\varepsilon'$ values are represented as a function of the temperature for samples with different compositions. As shown, the experimental values qualitatively follow Eq. 7 although $1/\varepsilon'$ achieves a non-zero value at the critical temperature. The intrinsic composite structure of semicrystalline polymers has been invoked to understand this effect [8, 6]. The order of magnitude of the constant A has been reported to be around $10^3 °C$ [11] which is consistent with the relatively high polarizability of these materials. At this point it is important to emphasize that the knowledge of morphological aspects of these copolymers might help, in future, to develop a theoretical framework capable of accounting for the experimental observations.

10 Molecular Dynamics: Incoherent Neutron Scattering

The study of molecular dynamics in polymers is of great interest because the structural changes of the crystal lattice are intimately related to the onset of molecular motions which generate a special type of dynamical disorder within the crystals [101–104]. In this final section, we present an experimental account of the molecular dynamics of copolymers with 60/40 and 80/20 VF_2/F_3E mole fraction composition using incoherent quasielastic neutron scattering.

We have performed a detailed analysis of the molecular motions in the three phases of these materials, namely ferroelectric, paraelectric, and in the melt [105]. The experiments were carried out using the cooled neutron backscattering spectrometer IN10 and the thermal neutron backscattering spectrometer IN13 at the Institut Laue-Langevin in Grenoble. The energy resolution of these spectrometers was set to 1 μeV and 9 μeV respectively. This allows one to investigate dynamics as slow as 10^{-9} s and 10^{-10} s on the instruments IN10 and IN11 respectively at the neutron research facilities at ILL.

Due to the large incoherent scattering cross-section of hydrogen ($\sigma = 79.90$ barns, 1 barn $= 10^{-28} m^2$) in comparison with the σ-values of carbon ($\sigma = 0$ barns) and fluorine ($\sigma = 0.0008$ barns) the mean contribution to the scattered intensity arises mainly from the hydrogen atoms in the molecule and overhelms any other influence. We have measured the double differential scattering cross-section $d^2\sigma/d\Omega dE$ which, neglecting the scattering of carbon and fluorine, gives directly the self scattering function $S_{inc}(Q, w)$. Measurements were carried out at several scattering vectors ($|Q| = 4\pi sen\Theta\lambda^{-1}$) covering the whole Q-range of the two instruments (between 0.2 and 2 Å^{-1} for IN10 and 0.2–5.3 Å^{-1} for IN13). For the quasielastic measurements, temperatures were selected in the ranges corresponding to the ferroelectric and paraelectric phases and above the melting point. The experimental resolution was obtained from measurements at 4 K, where it is assumed that there are no motions of the protons in the copolymers. The spectrum of a vanadium sample with the same shape and thickness of the polymer was also recorded and used in the standard way for calibration and conversion to $S(Q, w)$. In addition, the spectrum of an empty aluminium can was registered for background corrections. The fitting process includes a theoretical model which is convoluted with the resolution and then adjusted to the experimental spectrum in order to derive the free parameters of the model.

10.1 Elastic Backscattering

The elastic intensity was measured as a function of temperature during a heating and cooling cycle with the IN10 spectrometer. We used the so-called "fixed elastic window" method in which the monochromator and the analyzer are set to the same energy. Accordingly, only those neutrons which change their energy by an amount smaller than the energy resolution of the spectrometer are detected. At low temperatures where, apart from some vibrational motion, the proton is at rest, nearly all the scattered intensity passes through the fixed window. Here the decrease in the measured intensity is attributed to the proton Debye-Waller factor (DWF)

$$I \approx I_o \exp[-1/3\{Q^2\langle u^2\rangle\}] \qquad (8)$$

where I_o is the intensity measured at a very low temperature (4 K) and $\langle u^2\rangle$ the mean square vibration amplitude. If new motions are excited, the elastic line broadens up and only the center part within the fixed window can pass through the instrument and the observed intensity drops.

Figure 28 illustrates the variation of $\ln(I/I_o)$ for both copolymers at the same Q value, with increasing and decreasing temperature. For the 60/40 copolymer the scattering intensity follows an exponential decrease with Q showing three regions with increasingly larger slopes which correspond to the ferroelectric phase (T $\leq$ 340 K), to the paraelectric phase (340 $\leq$ T $\leq$ 427 K) and to the molten state (above 427 K). In the case of the 60/40 copolymer at the phase transitions, two hysteresis loops appear corresponding to the Curie transition and melting respectively, in agreement with the thermal hysteresis observed by means of other methods (see Sects. 7.1 and 9.3). For the 80/20 copolymer the intensity variation with T only shows one hysteresis loop. As we have seen in the DSC and microhardness experiments the Curie temperature of this copolymer is very close to the melting point ($T_{Curie} = 415$ K and $T_{melt} = 423$ K) and we just

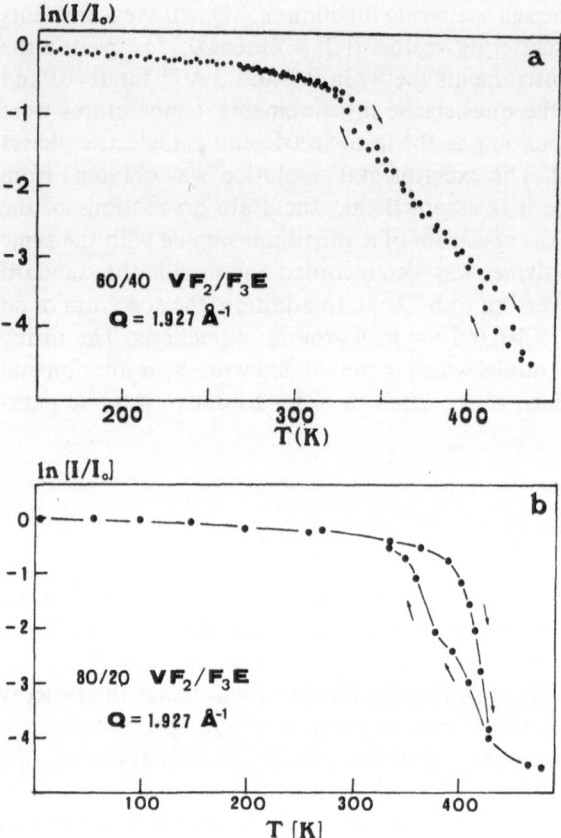

Fig. 28. Plot of $\ln(I/I_0)$ as a function of the temperature for the heating and cooling cycles of both polymers using the "fixed elastic window" method

observe a superposition of both hysteresis loops [62]. The elastic incoherent scattering of the copolymers markedly deviates near T_{Curie} from the usual DWF-behavior.

From the plot of $\ln(I/I_0)$ versus Q^2 the $\langle u^2 \rangle$ values for the temperatures below T_c were derived. At higher temperatures deviations from linear behavior occur at large Q. Figures 29a and 29b show the values of the mean square vibration amplitude, $\langle u^2 \rangle$, as a function of temperature for both copolymers. At low temperature the mean square displacement follows a nearly linear temperature dependence as expected for harmonic vibrations. A stronger and quasi-exponential temperature dependence sets in around $T = 250\ K$ for the 60/40 copolymer and $T = 230\ K$ for the 80/20 copolymer. It should be noted that the temperatures where a deviation from the harmonic behavior occurs corresponds to the glass transition in the case of both copolymers [6]. We can attribute this behaviour to the appearance of a new degree of freedom in this region. Similar

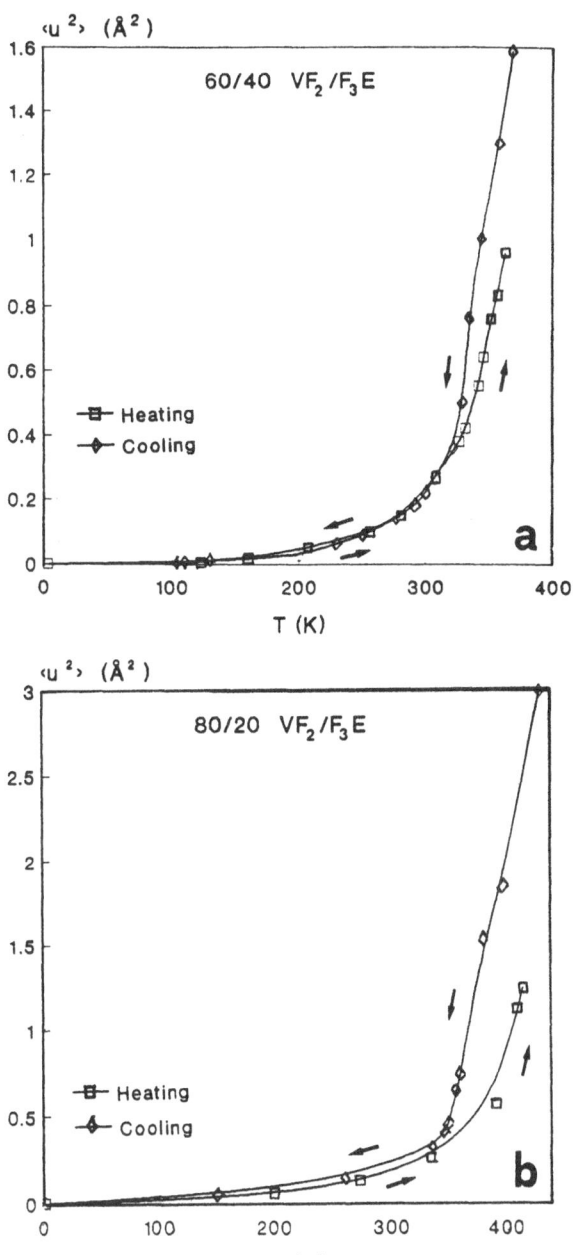

Fig. 29. Variation of the mean square vibration amplitude $\langle u^2 \rangle$ with temperature

effects have been detected in other amorphous and semicrystalline polymers [106].

10.2 Quasielastic Backscattering

We have investigated the quasielastic spectrum of the 60/40 copolymer at temperatures corresponding to the three phases. In the temperature range corresponding to the ferroelectric phase there is no quasielastic scattering within the dynamics range of the instrument. For $T \geq T_c$ a quasielastic broadening which increases with temperature was observed (see Fig. 30). When we approach the melting temperature the elastic component diminishes considerably. The experimental results were analyzed with a model that includes the sum of a delta and a Lorentzian function, i.e. we suppose that the motion is a simple relaxation process in a restricted volume. In this model we have two free parameters: the half-width at the half maximum (HWHM) of the Lorentzian and the intensity of the elastic component. Both parameters are functions of Q. We proceeded in the following way: we made the convolution of the function describing the model with the experimental resolution of the spectrometer. By comparing the convolution with the experimental results we derive the above-mentioned free parameters of the model.

Figure 31 shows the values of HWHM of the Lorentzian function versus the squared momentum transfer Q, at several temperatures. In this figure the data

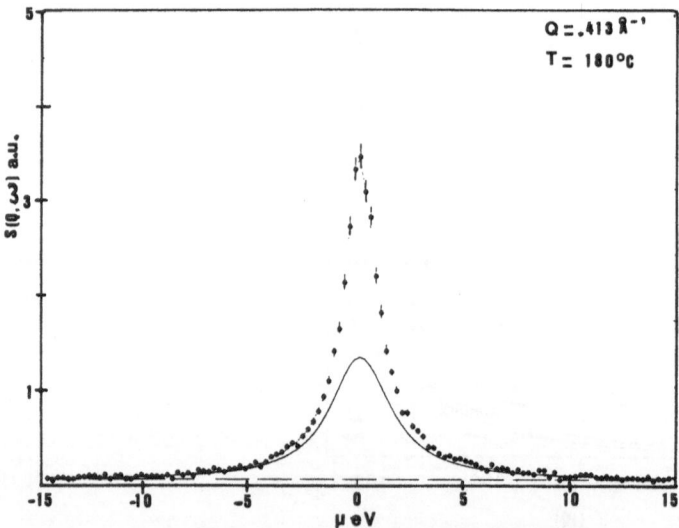

Fig. 30. Quasielastic broadening of the neutron scattering peak observed in the paraelectric phase for the copolymer 60/40. The *points* represented the experimental results with their statistical error bar. The *solid line* is the Lorentzian function obtained after the deconvolution process described in the text

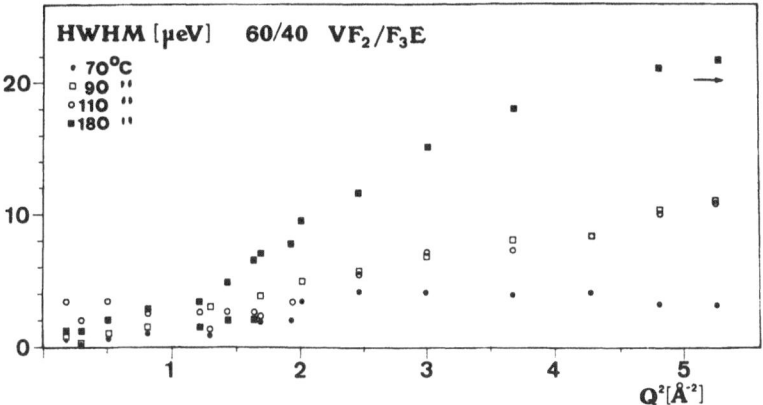

Fig. 31. HWHM of the Lorentzian function versus the momentum transfer Q at several temperatures in the paraelectric phase

obtained with the IN10 and IN13 spectrometers are presented. The two main characteristics of this dependence are the tendency to saturation at high Q values and the non-zero value in the limit $Q \to 0$. This behavior can be explained by a random jump-diffusion motion of hydrogen atoms of the chain molecule in a bounded media [107, 108]. The incoherent scattering function for such a model is given by:

$$S_{inc}(Q, w) = A_0(QL)\delta(w) + \sum_{n=1}^{\infty} A_n(QL)L_n(w, \Gamma) \tag{9}$$

where $A_0(QL)$ is the elastic incoherent structure factor (EISF): which is given by

$$A_0(QL) = 2[1 - \cos(QL)]/QL^2 = j_0^2[QL/2] \tag{10}$$

On the other hand, the HWHM of the Lorentzian functions $L_n(w, \Gamma)$ is expressed by:

$$\Gamma = 1/r [1 - \exp[-n^2 \pi r^2/2L^2]] \tag{11}$$

where L is equal or smaller than the distance between the neighboring chains, r^2 the mean square jump-length and τ the mean time between two successive jumps. The random-jump restricted diffusion model exhibits asymptotic behavior at low and high Q. At low Q, we are observing large distances, as boundary effects, and Γ approaches the value $\pi^2 r^2/2L^2\tau$. On the other hand, at large Q we are looking at short distances and approaches $1/\tau$.

The distance between neighboring chains L has been evaluated from the experimental EISF, and r and τ from the experimental asymptotic values of HWHM. Figure 32a shows that L increases strongly from 1 up to 4.8 Å with temperature while the jump parameter r varies only slightly (between 0.8 and 1.6 Å). At 340 K the r and L values are nearly the same indicating that at the onset of the Curie transition all the free volume available is occupied by the jump.

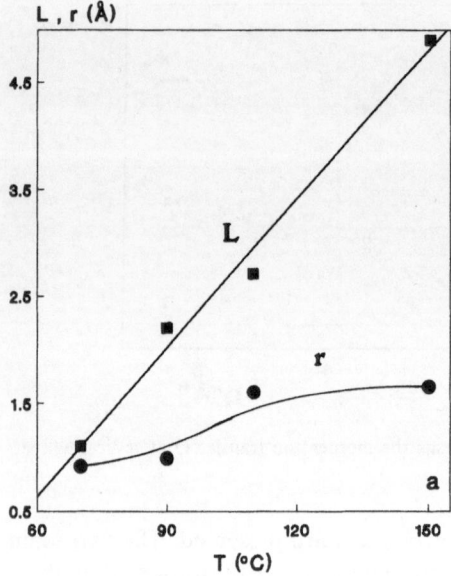

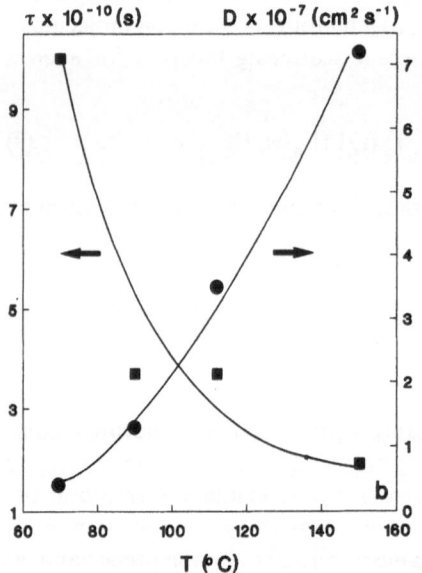

Fig. 32a, b. Temperature variation of: a. the geometrical parameters jump distance r and distance L between neighbouring chains. b. Residence time τ and diffusion constant D

Figure 32b shows the variation of the residence time between jumps τ with increasing temperature. Furthermore, we have calculated the value for the diffusion constant $D = r^2/2\tau$. One sees that D increases with temperature up to values of $D = 7 \times 10^{-7} \, cm^2 \, s^{-1}$ for $T = 430 \, K$. From the variation of D with temperature, assuming an activation law

$$D = D_0 \exp[-\Delta E_a/k_B T] \qquad (12)$$

where ΔE_a is the activation energy and k_B the Boltzmann constant, we obtain $\Delta E_a = 0.45$ meV.

11 Conclusions

In this review, recent structural results on para to ferroelectric phase transitions of vinylidene fluoride/trifluoroethylene copolymers derived from X-ray diffraction experiments and calorimetric studies have been discussed in some detail. Although this is a subject of some complexity due to the appearance of a mixture of polar and non-polar phases at room temperature, it has been possible to show that the systematic application of real time X-ray scattering techniques to these copolymers can provide unique information of considerable value. At high temperatures (paraelectric phase) the stacks of lamellae are found to consist of coherently diffracting blocks of a few hundred Å in width. On cooling from the paraelectric state through the Curie transition, the results obtained support the concept of a breakdown of the relatively large paraelectric crystals into smaller ferroelectric domains statistically mixed with paraelectric regions. It is shown that the remanent polarization P_r (measured at room temperature), after poling with high electrical fields, linearly increases with the fraction of ferroelectric crystals which depends upon VF_2 content. With increasing temperature, both the number of ferroelectric crystals and P_r gradually diminish. At T_c, the polar crystals are transformed into non-polar domains and, as a consequence, the remanent polarization vanishes.

The analysis of the real and imaginary part of the complex dielectric permittivity allows one to distinguish between the two main relaxation processes (α and β). The α-process is correlated to the transition from the ferro to the paraelectric phase and the β-process is attributed to segmental motions in the amorphous phase.

Microhardness (MH), has been shown to be a convenient additional technique to detect accurately the ferro to paraelectric phase changes in these copolymers. The increase of MH as a function of VF_2 polar sequences observed at room temperature is correlated with the contraction of the β-all-trans unit cell. On the other hand, the fast exponential decrease of MH with increasing temperature, observed above T_c, is similar to that obtained for glassy polymers above T_g and suggests the existence of a liquid crystalline state in the high temperature paraelectric phase. This phase is characterized by a disordered sequence of conformational isomers (tg^-, tg^+, tt) as discussed for Condis crystals [109].

Finally, the study of the protons of the polymer chain measured by incoherent neutron scattering allows the identification of two distinct types of motion: (a) a vibrational motion of the Debye-Waller type and (b) a slow jump-like diffusive motion of the whole chain confined within the volume restricted by the neighboring molecules. Characteristic values for the diffusion constant and residence time are reported.

Acknowledgments: Grateful acknowledgment is due to CICYT (Grant MAT-90-0795) Spain, and to the Internationales Büro Kernforschunganlage, Karlsruhe for the generous support of this investigation.

12 References

1. Seanor DA (ed) (1983) Electrical properties of polymers Academic, New York, Chap 5
2. Kawai H (1969) Jpn J Appl Phys 8: 975
3. Bergman Jr JG, Mc Tec JH, Grane GR (1971) Appl Phys Lett 18: 203
4. Sussner H, Dransfeld K (1978) J Polym Sci Polym Phys Ed 16: 529
5. Kepler RG, Anderson RA, Crit CRR (1980) Rev Solid State Mater Sci 9: 399
6. Yagi T, Tatemoto M, Sako JI (1980) Polym J 12: 209
7. Higashihata Y, Sako JI, Yagi T (1981) Ferroelectrics 32: 85
8. Furukawa T, Johnson GE, Bair HE, Tajitsu Y, Chiba A, Fukada E (1981) Ferroelectrics 32: 61
9. Yamada T, Ueda T, Kitayama T (1981) J Appl Phys 52: 948
10. Yamada T, Kitayama T (1981) J Appl Phys 52: 6859
11. Legrand JF, Lajzerowicz J, Berge B, Delzenne P, Macchi F, Bourgaux-Leonard C, Wicker A, Kruger JK (1988) Ferroelectrics 78: 151
12. Lovinger AJ (1983) Macromolecules 16: 1529
13. Green J, Rabolt JF (1987) Macromolecules 20: 457
14. Marand H, Stein DR (1989) Macromolecules 22: 444
15. Murata Y, Koizumi N (1989) Ferroelectrics 92: 47
16. Mathew SC, Scheinbein JI, Newmann BA (1984) J Appl Phys 56: 2419
17. Tajitsu T, Chiba A, Furukawa T, Date M, Fukoda E (1980) Appl Phys Lett 36: 286
18. Tashiro K, Takano K, Kobayashi M, Chatani Y, Tadokoro H (1981) Polymer 22: 1312
19. Dario P, de Rossi D, Giannotti G, Vivaldi F, Pinolli PC, (1984) Ferroelectrics 60: 199
20. Yamasaki H, Ohwaki J, Yamada T, Kitayama T (1981) Appl Phys Lett 39: 772
21. Betz R (1987) Ferroelectrics 75: 397
22. Vignolle JM, Chambost E, Micheron F (1989) Ferroelectrics 93: 5
23. Smith MU, Shanlov AA (1988) Ferroelectrics 76: 215
24. Cady WG (1946) Piezoelectricity, 1st edn, McGraw-Hill, chap 8
25. Gutierrez Monreal FJ, Mari CM (1987) Sensors and Actuators 12: 129
26. von Hippel A (1954) Dielectrics and waves, John Wiley, New York, p 95
27. Fatuzzo E, Merz WJ (1967) Selected topics in Solid State Physics: Ferroelectrics chap 1.2, North Holland
28. Broadhurst MG, Davis GT (1980) In: Sessler GM (ed) Topics in applied physics: Electrets, Springer, Berlin Heidelberg New York, chap 5
29. Wilson CW, Santee ER Jr (1965) J Polymer Sci C8: 97
30. Tonelli AE, Schilling FC, Cais, RE (1982) Macromolecules 15: 849
31. Lovinger AJ (1982) In: Bassett DC (ed) Developments in crystalline polymers, J Appl Sci London, vol 1, chap 5
32. Lovinger AJ (1983) Science 220: 1115
33. Kolda R, Lando JB (1975) J Macromol Sci B11: 21
34. Tashiro K, Takano K, Kobayashi M, Chatani Y, Tadokoro H (1984) Ferroelectrics 57: 297
35. Lovinger AJ, Cais RE (1984) Macromolecules 17: 1939
36. Galperin YL, Strogalin YV (1965) Vysokomol Soedin 7: 16
37. Lando JB, Doll WW (1968) J Macromol Sci Phys B2: 205
38. Lovinger AJ, Furukawa T, Davis GT, Broadhurst MG (1983) Polymer 24: 1225–1233
39. Lovinger AJ, Davis GT, Furukawa T, Broadhurst M (1982) Macromolecules 15: 323 and 329
40. Davis GT, Mc Kinnly JE, Broadhurst MG, Roth SC (1978) J Appl Phys 49: 4998

41. Dney-Aharon H, Taylor PL, Hopfinger AJ (1980) J Appl Phys 51: 5184
42. Lovinger AJ (1981) Macromolecules 14: 225
43. Yagi T, Tatemoto T (1979) Polym J 11: 429
44. Kepler RG, Anderson RA (1978) J Appl Phys 49: 1232
45. Weinhold S, Litt MH, Lando JB (1981) Macromolecules 14: 40
46. Lovinger AJ (1981) Macromolecules 14: 322
47. Lovinger AJ (1981) J Appl Phys 52: 5934
48. Nakagura K, Ishida Y (1973) J Polymer Sci Phys Ed 14: 2153
49. Gianotti G, Capizzi A, Zamboni V (1973) Chim Ind (Milan) 55: 501
50. Prest WM Jr, Luca DJ (1975) J Appl Phys 46: 4136
51. Lovinger AJ (1980) J Polym Sci Polym Phys Ed 18: 793
52. Lovinger AJ, Keith HD (1979) Macromolecules 12: 919
53. Lovinger AJ (1980) Polymer 21: 1317
54. Scheinbeim J, Nakafuku C, Newman BA, Pae KD (1979) J Appl Phys 50: 4399
55. Dunn PE, Carr SH (1989) MRS Bulletin 14: 22
56. Farmer BL, Hopfinger AJ, Lando JB (1972) J Appl Phys vol 43, 11: 4293
57. Furukawa T, Date M, Fukada E, Tajitsu Y, Chiba A (1980) Jap J Appl Phys 19: L109
58. Davis GT, Furukawa T, Lovinger AJ, Broadhurst MG (1982) Macromolecules 15: 329
59. Lovinger AJ, Furukawa T, Davis GT, Broadhurst MG (1983) Ferroelectrics 50: 227
60. Tashiro K, Takano K, Kobayashi M, Chatani Y, Tadokoro H (1984) Polymer 25: 195
61. Davis GT, Broadhurst MG, Lovinger AJ, Furukawa T (1984) Ferroelectrics 57: 73
62. Lopez-Cabarcos E, Gonzalez Arche A, Balta-Calleja FJ, Zachmann HG (1988) Makromol Chem Macromol Symp 20/21: 193
63. Fernandez MV, Suzuki A, Chiba A (1987) Macromolecules 20: 1806
64. Koizumi N, Haikawa N, Habuka H (1984) Ferroelectrics 57: 99
65. Lovinger AJ (1985) Macromolecules 18: 910
66. Macchi F, Daudin B, Legrand JF (1990) Nuclear Instruments and Methods in Phys Research B46: 324
67. Macchi F, Daudin B, Hillairet J, Lauzier J, N'Goma JB, Cavaille JY, Legrand JF (1990) Nuclear Instruments and Methods in Phys Research B46: 334
68. Green JS, Farmer BL, Rabolt JF (1986) J Appl Phys 60(8): 2690
69. Tashiro K, Nishimura S, Kobayashi M (1988) Macromolecules 21: 2463
70. Tanaka H, Yukawa H, Nishi T (1988) Macromolecules 21: 2469
71. Bongianni W (1990) Ferroelectrics 103: 57
72. Delzenne P (1986) PhD Thesis, Univ of Grenoble France
73. Tashiro K, Kobayashi M (1986) Polymer 27: 667
74. Tashiro K, Kobayashi M (1988) Polymer 29: 426
75. Legrand JF (1989) Ferroelectrics 91: 303
76. Martinez Salazar J, Canalda JC, Lopez Cabarcos E, Balta Calleja FJ (1988) Colloid & Polymer Sci 66: 41
77. Legrand JF, Delzenne P, Lajzerowicz J (1986) Proc of the International Symosium on the Applications of Ferroelectrics (ISAF86). Ed by V Wood. Lehigh University Bethlehem Pa
78. Lopez Cabarcos E, Gonzalez Arche A, Martinez Salazar J, Balta Calleja FJ (1988) Integration of Fundamental Polymer Science and Technology, Elsevier, London, New York 2: 193
79. Lopez-Cabarcos E, Gonzalez Arche A, Balta-Calleja FJ, Bösecke P, Röber S, Bark M, Zachmann HG (1991) Polymer 32: 3097
80. Legrand JF (1989) Ferroelectrics 91: 303
81. Gonzalez Arche A (1990) Doctoral Thesis, Univ Complutense de Madrid
82. Tajitsu Y, Ogura H, Chiba A, Furukawa T (1987) Jpn J Appl Phys 26: 554
83. Koga K, Ohigashi H (1986) J Appl Phys 59: 2142
84. Farago B (1989) ILL Annual Report
85. Baltá Calleja FJ (1985) Adv Polym Sci 66: 117
86. Baltá Calleja FJ, Kilian HG (1985) Colloid & Polym Sci 263: 697
87. Baltá Calleja FJ, Kilian HG (1988) Colloid & Polym Sci 266: 29
88. Baltá Calleja FJ, Martinez Salazar J, Asano T (1988) J Mat Sci Lett 7: 165
89. Kasap SO, Yannacopoulos S, Gundappa P (1989) J Non-Cryst Solids 111: 82
90. Ania F, Martinez Salazar J, Baltá Calleja FJ (1989) J Mater Sci 24: 2934
91. Baltá Calleja FJ, Santa Cruz C, Bayer RK, Kilian HG (1990) Colloid & Polym Sci 268: 440
92. Baltá Calleja FJ, Santa Cruz C, Sawatari C, Asano T (1990) Macromolecules 23: 5352
93. Hedvig P (1977) Dielectric Spectroscopy of Polymers. Adam Hilger, Bristol

94. Kremer F, Vallerien SV, Zentel R (1990) Adv Mater 2: No 3, 145
95. Blythe AR (1979) Electrical Properties of Polymers, Cambridge Univ. Press, Cambridge
96. Furukawa T, Johnson GE (1981) J Appl Phys 52(2): 940
97. Hahn B, Wendorff J, Yoon D (1985) Macromolecules 18: 718
98. Tashiro K, Takamoto K, Kobayashi M, Chatani Y, Tadokoro H (1981) Polymer Communication 22: 1312
99. Havriliak S, Negami S (1967) Polymer 8: 161
100. Kittel C (1986) Introduction to Solid State Physics, John Wiley & Sons Inc, New York
101. Petzet J, Pacesová S, Kamba S, Legrand JF, Kozlov GV, Volkov A (1988) Ferroelectrics 80: 205
102. Legrand JF, Shuele PJ, Schmit VH, Minier M (1985) Polymer 26: 1683
103. Hirchinger J, Meurer B, Weill G (1987) Polymer 28: 721
104. Lopez-Cabarcos E, Gonzalez Arche A, Batallan F, Frick B (1989) Physica B 156/157: 423
105. Arche AG, Batallán F, Frick B, Baltá Calleja FJ, López Cabarcos E (1991) Spanish Scientific Research Using Neutron Scattering Techniques. Univ. Cantabria, p. 197
106. Frick B, Richter D (1989) Springer Proceedings in Physics 37/38
107. Hall PL, Ross DK (1981) Molecular Physics 42: 673
108. Bee M (1988) Quasielastic Neutron Scattering. Adam Hilger, Bristol, England
109. Wunderlich B (1990) Thermal Analysis, Academic, Boston, p 31

Received November 11, 1991

Packing of Chain Segments: A Method for Describing X-Ray Patterns of Crystalline, Liquid Crystalline and Non-Crystalline Polymers

T. Pieper, H.-G. Kilian
University of Ulm, Department of Experimental Physics,
Albert-Einstein-Allee 11, D-7900 Ulm, FRG

A general principle of chain segment packing is presented which covers a wide range of distance correlations occurring in macromolecular systems. The variety of local motions and defects in chain segments results in a delocalized, cylindrically symmetric electronic distribution along the chain. Thus, the molecules can be substituted for cylinder segments containing these conformationally averaged chain segments. The length of these cylinders is completely determined by the atom-to-atom distance range of the molecular autocorrelation function. The conformation elements obtained by appropriate statistical calculations were packed in the lateral direction by applying well-known principles from one-dimensional fluid models obeying the correct values of the mean site-to-site distance, distance fluctuations, exclusion volume, and macroscopic density.

Short range order in liquid-like systems as well as long range order in crystalline domains are reflected in WAXS-patterns very clearly. Some examples of calculated X-ray patterns from PTFE (Phase I), a smectic LC-phase and even a PE melt, show that our model covers a wide range of macromolecular structures running the whole scale from crystalline systems over mesophases up to polymer melts. The range of intra- and intermolecular order can be estimated fairly well with the help of density correlation functions.

Advances in Polymer Science, Vol. 108
© Springer-Verlag Berlin Heidelberg 1993

List of Abbreviations and Symbols

$\vec{Q}$	scattering vector
λ	wavelength
θ	Bragg angle
2θ	scattering angle
$\hat{\rho}(\vec{r})$	density correlation function (DCF)
$G(r)$	radial density difference function (RDDF)
$g_{NN}(r)$	distance distribution of next neighbours
$g(r)$	radial distance distribution
$\hat{\rho}_{auto}(\vec{r})$	auto-correlation function
$\langle \hat{\rho}_{auto}(\vec{r}) \rangle$	mean auto-correlation function
ρ_0	mean electron density
$\langle \rho(\vec{r}) \rangle$	mean molecular electron density
n_0	mean number density
η_L	lateral packing density (number of cylinders per unit area)
n_{CH_2}	number density of CH_2 units
n_{cyl}	number of CH_2 units in correlation cylinder
$n_{packing}, n_{chain}$	radial and axial CH_2 number density, respectively
F_k, F_l^*	molecular structure amplitudes
$I_{intra, L}$	isotropic average of mean intramolecular structure factor in laboratory system
$I_{inter, L}$	isotropic average of mean intermolecular structure factor in laboratory system
f_i, f_j	form factors of atoms i and j
$\langle |F|^2 \rangle$	mean molecular structure factor
$\langle F \rangle$	mean molecular structure amplitude
RISA	Rotational Isomeric State Approximation
p_j	statistical weight of molecular conformation j
T	transition matrix
r_H	hard core diameter
$\langle r \rangle$	mean distance
R_e^2	electronic radius of gyration
g	distance fluctuation parameter, $g = \sqrt{\dfrac{\langle r^2 \rangle}{\langle r \rangle^2} - 1}$
r, z	radial and axial component of $\vec{r}$
q_r, q_z or R, Z	radial and axial component of $\vec{Q}$
q	$q = 2\pi \dfrac{2\sin\theta}{\lambda}$
$\vec{r}_{kl}$	distance vector between atoms k and l
$\langle \ldots \rangle$	space average
J_n	Bessel function of order n
$g_j(\varphi)$	angular distribution of atom j

a_{nj}	Fourier expansion of $g_j(\varphi)$
$\bar{M}_w$	molecular weight
$i(q)$	reduced scattering intensity
N_A	Avogadro number
MC	Monte-Carlo
PE	polyethylene
PTFE	polytetrafluorethylene

1 Introduction

The molecular short range order of distorted polymer structures such as polymer melts and amorphous polymers plays an important role for the evaluation of polymer microstructure models. It can be investigated with several physical methods. The determination of the stress optical coefficient [1], depolarized light scattering [2] and magnetic birefringence [2], for example, provide information about segment orientation within polymer chains. It was found that the range of intramolecular orientation correlation in rubber networks is very small restricting to a typical axial correlation length of 5–10 Å [1].

On the other hand, WAXS measurements of PE melt clearly indicate a range of intermolecular distance correlations of about 25 Å [3]. Together with the relatively high density of polymer melts, the fact that the first interchain halo in WAXS patterns of oriented amorphous polymers tends to lie in the equatorial direction and the relatively high WAXS intensity of the interchain halo support the idea of parallel chain segments on the short range scale.

As the WAXS pattern of a polymer melt shows all the characteristics of liquid-like scattering, it is obvious that molecular models be constructed with liquid-like radial distribution functions (RDF). The essential topological structure element of a simple fluid is given by the fact that the constituents cannot interpenetrate each other, leading to the representation by an effective atomic/molecular exclusion volume. A number of liquid properties can be explained in terms of effective hard core exclusion volumes. To apply this concept to polymers, it is necessary to segment the polymer chains into parts of appropriate length and electron density taking into account all the possible local conformations and orientations. The length of those segments is completely determined by the atom-to-atom distance range of the molecular autocorrelation function. This approach leads us to the concept of intramolecular correlation cylinders with delocalized electronic density as a consequence of space averaging.

Our structure model of the short range order in polymer melts combines the geometrical concepts of chain segmentation into 'X-ray-equivalent structure units', the transformation onto 'effective segment exclusion volumes' and their lateral packing in a natural manner. It will be shown that this approach gives correct values for lateral and axial distance correlations as well as a lower limit of the macroscopic density. Also, WAXS pattern and RDF are represented very well covering a large range in reciprocal and real space, respectively.

Additionally, the abilities of our model are exemplified by calculating X-ray patterns from PTFE (Phase I) and a smectic LC-phase of a polyester using the concepts of chain segmentation and space-averaged electron delocalization mentioned above. Thus it is verified that our model describes a wide range of different macromolecular structures covering the whole scale from crystals over mesophases up to polymer melts.

A crucial aspect of our approach is based on the idea that segmental rotation is activated thermally as the simplest mode of collective motions. This kind of first order distortion substantiated by oscillatory or diffusive segmental rotation is possible in a crystal lattice as well. This must not necessarily affect the ideal periodic structure but can be discussed in terms of the molecular structure factor.

Collective segmental rotation is considered to be the natural form gaining conformational entropy as the essential factor for stabilizing mesophases or liquid-crystalline structures. Despite the approximate character of our conception it should be possible to identify the significant characteristics of the intermolecular segmental arrangement in mesophases, in liquid crystals or in a polymer melt.

2 A Model of the Short Range Order in Distorted Polymer Structures

The chain conformation of macromolecules can always be represented by a phantom chain consisting of more or less extended segments which are linked freely ('Kuhn-segment'). Even if the equilibrium conformations of a free chain are discussed we are led to define "longitudinal intramolecular orientation correlations" in terms of the persistence length. This representative statistical segment produces the constructive intramolecular interference phenomena in the WAXS-pattern. The length of those segments is, of course, also influenced by their lateral packing arrangement, for example in a lattice or a liquid-like cluster.

The partition of molecular distance correlations into intra- and intermolecular contributions allows us to interpret these correlations in terms of a simple geometrical model. By this means, we are able to elicit structural units as for example segment-clusters that include intermolecular interference phenomena. These clusters are the "primary structure units" which we call "monodomains". These natural units characterize the basic symmetry of the whole structure. If we keep in mind this basic symmetry, we can construct our structure model from a molecular level up to the level of the monodomain treating intra- and intermolecular correlations independently. If we do so, every X-ray pattern can be represented by accounting for the orientation distribution of these monodomains.

Let us proceed to define appropriate coordinate systems (see Fig. 1 and Table 1). We use four different levels of coordinate frames in the averaging procedures of our model construction scheme to describe the molecular arrangement in the scattering volume. On the first level, System M describes the real structure of a segment, whereby the origin is fixed to a well-defined molecular unit. System MC (second level) is defined in cylindrical coordinates with the symmetry axis given by the long axis of the rod-like "molecular

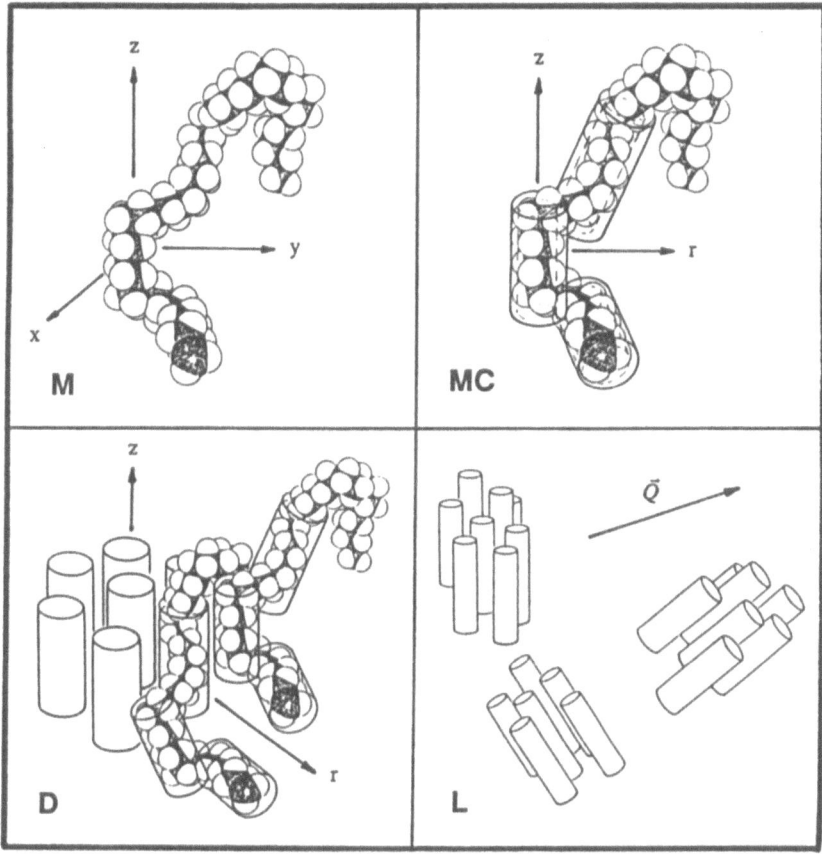

Fig. 1. Definition of body-fixed and space-fixed coordinate systems. Refer to Table 1 and to the text for a detailed explanation

Table 1. Definition of coordinate frames

Coordinate system	Description	Components of direct space vector	Components of reciprocal space vector
system M	Cartesian coordinate frame fixed to molecule	x, y, z	X, Y, Z
system MC	Cylindrical coordinate frame fixed to molecular cylinder defined by average molecular conformation	r, z	q_r, q_z R, Z
system D	Cylindrical coordinate frame fixed to z-axis of monodomain consisting of lateral arrangement of molecular cylinders MC	r, z	q_r, q_z R, Z
system L	Laboratory coordinate frame defined by the scattering vector		Q, q

cylinder" (MC) which results from the appropriate space average taken from all possible molecular conformations and angular orientations around the long axis. The molecular cylinder serves as a kind of substitute of the average molecular conformation. The third level concerns the lateral arrangement of these molecular cylinders within a monodomain D the size of which is given by the short range order of the molecular packing. Depending on the cluster structure the symmetry of a monodomain (D) may differ from the symmetry of the molecular cylinder (MC). At last, on the fourth level L (laboratory system) we assume that the monodomains are randomly oriented with respect to the scattering vector $\vec{Q}$ that is defined by the experimental setup.

2.1 Calculation of the X-ray Structure Factor

The calculation of the X-ray structure factor of distorted polymer structures should be based on a strategy of where to place approximations. As shown by the literature this is a difficult task.

The coherent X-ray scattering component of a given molecular configuration measured in system L (laboratory system) is written as

$$I_{tot, L}(\vec{Q}) = \sum_k \sum_l \langle F_k F_l^* \exp(i\vec{Q}\,\vec{r}_{kl}) \rangle \tag{1}$$

with $\vec{Q}$:   scattering vector in laboratory system coordinates (L);

$$|\vec{Q}| = 2\pi \frac{2\sin\theta}{\lambda}$$

F_k: structure amplitude of molecule k
F_l^*:   complex conjugate of F_l
$\langle \ldots \rangle$:   space average with respect to all possible molecular conformations, distance fluctuations, and orientations.

The correct calculation of the average in Eq. (1) is very extensive. For various reasons it appears to be justifiable to separate the average in Eq. (1) into average values of the molecular structure amplitudes and the phase term, respectively.

The calculation of the monodomain structure factor requires several averaging procedures to account for all possible molecular conformations and orientation configurations. For reasons of clarity we have labelled each of these averaging procedures according to Table 2.

(1) The first point implies a certain autonomy of chain segments. The simplest approach is a single chain approximation where intramolecular conformations are considered as being achieved independently of the conformation of adjacent chain segments. It is well established from experiment that the radius of gyration does not change dramatically between θ-solvent and melt [4, 5]. Therefore it is obvious to use the conformation potential of isolated chains as a basis for the chain segmentation in first approximation.

(2) Intrinsic symmetry conditions are likely to control vibrations, torsions and rotations of molecular groups within the segments involved. They induce,

Table 2. Labels of the averaging procedures

---

(1) molecular conformations
(2) torsions, oscillations and rotations of chain segments
(3) distance distribution of molecular centers
(4) rotation around z-axis of the monodomain
(5) azimuthal tilting of monodomain z-axis with regard to the space-fixed coordinate frame

---

nevertheless, modifications of the same type within segments in a different conformation.

(3) A delicate problem is the "lateral distance fluctuations" of segments within a monodomain of interest. An adequate but simplifying assumption is to define these fluctuations with regard to a symmetry axis of the segments. One encounters substantial complications in detail unless the axes are assumed to be parallel. The symmetry axes of the segments must be defined dependent on their actual conformation to make this an acceptable approximation. This results in representing each of the segments by a fictitious cylinder enveloping all the molecular units involved. Packing these cylinders in a parallel arrangement to form a monodomain, the fluctuation of distances defines its actual structure. Every description of these fluctuations can therefore be done in a plane perpendicular to the axis in the segment domain. This is a proper description that allows for longitudinal shifts of segments which are automatically accounted for if the conformation of each segment is related to a unit of reference which is selected at random. It is a crucial point that some molecular units of segments may interpenetrate the fictitious cylinders. But this should not, on average, substantially disturb the axial symmetry within each domain which must therefore be taken as a feature of the topological structure is distorted polymer systems, even in polymer melts.

We took care in describing our approach more extensively. It shows clearly that here is a possibility to study how inter- and intramolecular properties in polymer systems with very different structures (mesophases, smectic polymers, melt) are interrelated.

(4) The concept of defining monodomains in the way shown above includes (with the exception of pathologic cases) that, in the assembly of monodomains, each orientation occurs with the same probability. So it should be possible to construct a rotationally averaged representative monodomain.

(5) This last transformation is a straight-forward operation as long as we are dealing with macroscopically isotropic systems. In the case of textures the orientational average has to be modified according to a procedure given by Deas [6].

The double sum in Eq. (1) extends over all pairs of molecules k and l. It turns out to be useful to separate Eq. (1) into intra- (k = l) and intermolecular (k ≠ l) components giving

$$I_{tot,L}(\vec{Q}) = \sum_k \langle |F_k|^2 \rangle + \sum_{k \neq l} \sum \langle F_k F_l^* \exp(i \vec{Q} \vec{r}_{kl}) \rangle$$

$$= I_{intra,L} + I_{inter,L} \qquad (2)$$

The first term on the right hand side gives the intramolecular contribution in the laboratory system which depends on conformation and orientation of the relevant single segments; that means it depends on the mean intramolecular structure. The second term requires knowledge about the intermolecular structure within the monodomains.

2.2 Intramolecular Scattering Contributions $I_{intra,L}$

The intramolecular structure factor $I_{intra,M}$ given in body-fixed coordinates (M) can be written in terms of atomic scattering factors and the atomic distance distribution, yielding

$$I_{intra,M} = |F|^2_M = \sum_k \sum_l f_k f_l \exp(i\vec{r}_{kl}\vec{Q}) \tag{3}$$

where k, l   atom indexes
f_k, f_l atomic scattering factors of atom k and l
$\vec{r}_{kl}$ distance vector between atoms k and l
$\vec{Q}$ scattering vector in coordinates of system M.

In order to obtain the isotropic averaged intramolecular scattering contribution $I_{intra,L}$, we have to transform the above into our laboratory system L performing the appropriate averaging procedures. To do so, we rewrite $\vec{r}_{kl}$ and $\vec{Q}$ in spherical coordinates:

$$\vec{r}_{kl} = r_{kl}\begin{pmatrix}\cos\varphi\sin\vartheta\\\sin\varphi\sin\vartheta\\\cos\vartheta\end{pmatrix}, \quad \vec{Q} = q\begin{pmatrix}0\\0\\1\end{pmatrix} \tag{4}$$

with φ, ϑ spherical angles,
and $r_{kl} = |\vec{r}_{kl}|$, $q = |\vec{Q}|$.

The special choice of $\vec{Q}$ in the z-direction can be made without any restriction of generality. The isotropic average of the intramolecular scattering contribution is obtained by a spherical integration

$$I_{intra,L} = \langle|F|^2\rangle = \frac{1}{4\pi}\int_\Omega |F|^2_M d\Omega = \frac{1}{4\pi}\int_0^{2\pi}\int_0^\pi |F|^2_M d\varphi\sin\vartheta\, d\vartheta$$

$$= \frac{1}{4\pi}\sum_k\sum_l f_k f_l \int_0^{2\pi}\int_0^\pi \exp(ir_{kl}q\cos\vartheta)\,d\varphi\sin\vartheta\,d\vartheta$$

$$= \frac{1}{2}\sum_k\sum_l f_k f_l \int_0^\pi \exp(ir_{kl}q\cos\vartheta)\sin\vartheta\,d\vartheta \tag{5}$$

By replacing $\sin\vartheta\,d\vartheta$ for $-d(\cos\vartheta)$, we finally arrive at the common Debye formula

$$I_{intra,L} = \langle|F|^2\rangle = \sum_k\sum_l f_k f_l \frac{\sin(r_{kl}q)}{r_{kl}q}$$

$$= \sum_k f_k^2 + 2\sum_{k>l}\sum f_k f_l \frac{\sin(r_{kl}q)}{r_{kl}q} \tag{6}$$

For rigid distances $\hat{r}_{kl}$, Eq. (6) gives the gas-like scattering of an isotropic assembly of segments. It is important to note that only the lengths $|\hat{r}_{kl}|$ enter into the equation independent of their actual orientation within the segments. So, strictly speaking, all structures with the same atomic distance distribution are equivalent. Every interpretation of this term needs – even in the limits of rigid distances – a pre-knowledge of the actual chain conformation.

Accounting for molecular conformations or torsional and rotational chain dynamics, it is more useful to calculate the mean intramolecular structure factor in terms of density correlation functions (DCF). The structure factor results simply from a Fourier transform of the corresponding DCF.

In the intramolecular case, the DCF as the autocorrelation function is given by

$$\hat{\rho}_{\text{auto}}(\tilde{r}) = \int\limits_{\text{molecule}} \rho(\tilde{r}')\rho(\tilde{r} + \tilde{r}')d^3r' \tag{7}$$

The molecular structure factor results from the Fourier transform of Eq. (7), yielding

$$|F|^2 = \int \hat{\rho}_{\text{auto}}(\tilde{r})\exp(i\tilde{r}\vec{Q})d^3r \tag{8}$$

where the integral extends over the whole molecule.

To account for the effects of conformational and torsional changes, for example those caused by rotational isomers, one has to introduce the properly weighted conformationally averaged electron density $\langle \rho(\tilde{r}) \rangle$:

$$\langle \rho(\tilde{r}) \rangle = \sum_j p_j \rho_j(\tilde{r}) \tag{9}$$

with

   j: index to label the molecular conformation

   p_j: relative frequency of molecular conformation j; $\sum_j p_j = 1$

As a consequence of the single chain approximation the conformation statistics of any two chains are independent of each other. Correspondingly, the mean autocorrelation function is given by

$$\langle \hat{\rho}_{\text{auto}}(\tilde{r}) \rangle = \sum_j p_j \left(\int\limits_{\text{molecule}} \rho_j(\tilde{r}')\rho_j(\tilde{r} + \tilde{r}')d^3r' \right) \tag{10}$$

and the Fourier transform yields to the averaged molecular structure factor

$$\langle |F|^2 \rangle = \int \langle \hat{\rho}_{\text{Auto}}(\tilde{r}) \rangle \exp(i\tilde{r}\vec{Q})d^3r \tag{11}$$

In the isotropic case the orientation average is performed analogous to Eqs. (3)–(6). The result is

$$\langle I_{\text{intra,L}} \rangle = \sum_j p_j \left(\sum_k \sum_l f_k f_l \frac{\sin(r_{kl}^{(j)} q)}{r_{kl}^{(j)} q} \right) \tag{12}$$

where $r_{kl}^{(j)}$ denotes the distance between atoms k and l of the molecule in conformation j. The conformational weighting is given by p_j as above. If rotational isomers are present the calculation of the statistical weights p_j with regard to molecular conformations is performed by the method of the Rotational Isomeric State Approximation (RISA) which is briefly outlined in Sect. 2.4.

2.3 Intermolecular Scattering Contributions $I_{inter, L}$

Now we must look at the consequences of the averaging procedures (1)–(5) with regard to the monodomain structure factor (see Table 2). First we calculate the intermolecular part $I_{inter, D}$ of the structure factor of a single domain and apply the averaging procedures (1)–(4) to it:

$$I_{inter, D} = \sum_{k \neq l}\sum \langle F_k F_l^* \exp(i \vec{Q} \vec{r}_{kl}) \rangle_{1234} \tag{13}$$

$I_{inter, D}$ describes the intermolecular scattering function in a coordinate frame fixed to the z-axis of the monodomain. Focussing our interest on liquid-like structures we may assume that distance fluctuations of the molecular sites do not affect the conformation statistics of segments. Using this single segment approach, the averaging procedures (1) and (2) can be carried out independent of (3) and influence only $F_k F_l^*$. The averaging procedure (4) (rotation around the z-axis of the monodomain) then leads to a representative monodomain with cylindrical symmetry. The molecular structure amplitudes are separated from the monodomain phase term according to

$$I_{inter, D} = \sum_{k \neq l}\sum \langle F_k F_l^* \rangle_{12} \cdot \langle \exp(i \vec{Q} \vec{r}_{kl}) \rangle_{34}. \tag{14}$$

Next, we examine the term $\langle F_k F_l^* \rangle_{12}$. In a "gas-like" single segment approximation, this term can be replaced by $\langle F_k \rangle_{12} \langle F_l^* \rangle_{12}$. The molecular conformation statistics are independent of each other. This might be due to the fact that in the absence of a three-dimensional lattice-potential, nematic shifts of neighboring segments are very likely to occur. In this approximation the configuration does not depend on which individual pair of molecules k, l is picked out. The molecular structure factor is independent of the indexes k and l. Hence $I_{inter, D}$ can be written as

$$I_{inter, D} = |\langle F \rangle|^2 \cdot \sum_{k \neq l}\sum \langle \exp(i \vec{Q} \vec{r}_{kl}) \rangle_{34}. \tag{15}$$

The double sum in Eq. (15) can now be expressed by a cylindrical density correlation function n(r) which counts the number of molecular sites lying in a circular shell of thickness dr with distance r to a given point of reference on the symmetry axis of the monodomain. This RDCF contains the lateral distance statistics of segments within a domain – procedure (3), as well as the average of the domain ensemble – procedure (4). Due to the cylindrical symmetry we finally obtain

$$I_{inter,D} = |\langle F \rangle|^2 \cdot \int_0^\infty n(r) J_0(rR) 2\pi r \, dr$$

$$= |\langle F \rangle|^2 \cdot (S - 1). \tag{16}$$

S is the cylindrical symmetric configuration structure factor given by

$$S = 1 + \int_0^\infty n(r) J_0(rR) 2\pi r \, dr = 1 + n_0 \int_0^\infty g(r) J_0(rR) 2\pi r \, dr \tag{17}$$

with $g(r) = \dfrac{n(r)}{n_0}$: RDCF normalized to 1

and J_0 Besselfunction of order zero
$\quad n_0$ mean density, $r \to \infty$
$\quad R$ radial component of scattering vector $\vec{Q}$.

To carry out the last averaging procedure (5), we have to integrate the monodomain structure factor $I_{inter,D}$ in system D with regard to the laboratory system L given in polar coordinates formulated as

$$I_{inter,L} = \int_0^{\pi/2} I_{inter,D} \sin \alpha \, d\alpha. \tag{18}$$

α is the angle between the z-axis of monodomain D and an arbitrary polar axis in system L.

Now we are able to express the whole monodomain structure factor in terms of the intra and intermolecular structure factors already calculated:

$$I_{tot,L} = \sum_k \sum_l f_k f_l \frac{\sin(r_{kl} q)}{r_{kl} q} + \langle |\langle F \rangle|^2 \cdot (S - 1) \rangle_5$$

$$= I_{intra,L} + I_{inter,L}, \tag{19}$$

where the second term with index 5 denotes the final orientational average with respect to the laboratory system L. The calculation of the WAXS-pattern of models with an orientational distribution of the domains requires modification of the very last averaging procedure. Both the intramolecular and the intermolecular scattering contributions are affected. A convenient mathematical description was given by Deas [6].

2.4 Intramolecular Conformations and Dynamics

Small oscillatory torsions of molecular units within the segment produce a delocalization of the electronic density according to the "Debye-Waller effect". In a very distorted structure as in the melt its smooth intensity distribution is not affected to a measurable extent. Rotational isomers, on the other hand, induce remarkable changes of the r_{kl} involved. Moreover, it is easily shown that the Boltzmann weighting factor of the molecular conformation ensemble also has a substantial influence on the r_{kl}. It is strictly not possible to calculate the

WAXS-pattern of a polymer melt if these "Boltzmann-weighted distance fluctuations" are not accounted for. They characterize a relevant feature of polymer melt structures. We treat this in terms of the Rotational Isomeric State Approximation (RISA) discussed in the next section.

2.5 Rotational Isomeric State Approximation

First we will explain how to generate conformations of PE chain segments with the use of the RISA. We adopted the method used by Mitchell et al. [7] and added an important modification. The transition matrix given by Abe et al. [8] at $T = 400 \text{ K}$

$$
T = \begin{pmatrix} p_{tt} & p_{tg^+} & p_{tg^-} \\ p_{g^+t} & p_{g^+g^+} & p_{g^+g^-} \\ p_{g^-t} & p_{g^-g^+} & p_{g^-g^-} \end{pmatrix} = \begin{pmatrix} 0.54 & 0.23 & 0.23 \\ 0.683 & 0.291 & 0.026 \\ 0.683 & 0.026 & 0.291 \end{pmatrix} \tag{20}
$$

defines the combined probabilities that bond $(i - 1)$ is given by a particular rotational isomeric state $[t, g^+, g^-]$ and bond (i) by another particular isomeric state. The probability of g^+g^-- and g^-g^+-sequences is very low which is obvious from the corresponding transition probabilities in T. This is known as the 'pentane effect'. Additionally, the values of the rotation angles are only taken as mean values. In reality, these angles fluctuate due to thermal motion. We accounted for this effect by adding an adequate Gaussian distribution to the distribution of the rotation angles. With the help of appropriate computer programs, a set of molecular conformations with appreciable statistical weight can be generated. Furthermore we calculate the energy difference ΔE to the all-trans conformation for each conformation (j) and the corresponding Boltzmann-weights $p_j = \exp(-\Delta E/k_B T)$ (see Eq. (12)). For this purpose we used the conformational potential from a semi-empirical set of potential functions introduced by Haegele [9, 10]. Then we took the average of a sufficient number of simulation runs obtaining the mean atomic and electronic density functions and the mean autocorrelation function of the chain ensemble. The calculation of the intramolecular structure factor is easily performed by a Fourier transformation of the ACF. The results of a simulation run performed over 500 chain conformations comprising $C_{36}H_{72}$-segments are shown in Figs. 2 and 3.

The autocorrelation function (Fig. 2) shows clearly that a representative chain segment embracing about 5–7 CH_2-units should be sufficiently long to describe the intramolecular interference modulation in the WAXS-pattern of a PE-melt completely up to distances of approx. 30 Å.

We would like to stress the point that this is the consequence of accounting for the Boltzmann distribution by which a relevant statistical fluctuation of the r_{kl}'s is coming into play. If the thermal fluctuations of the rotation angles are not accounted for, the orientation correlation of the bond vectors along the chain would be much higher, resulting in a longer persistence length.

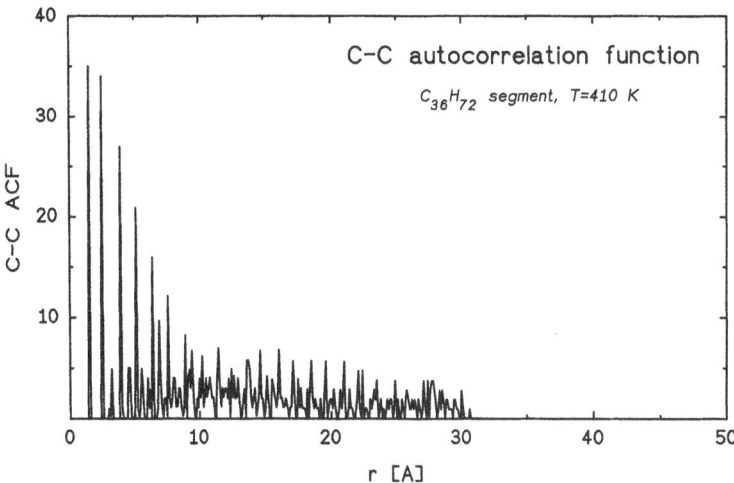

Fig. 2. Autocorrelation function of C...C distances

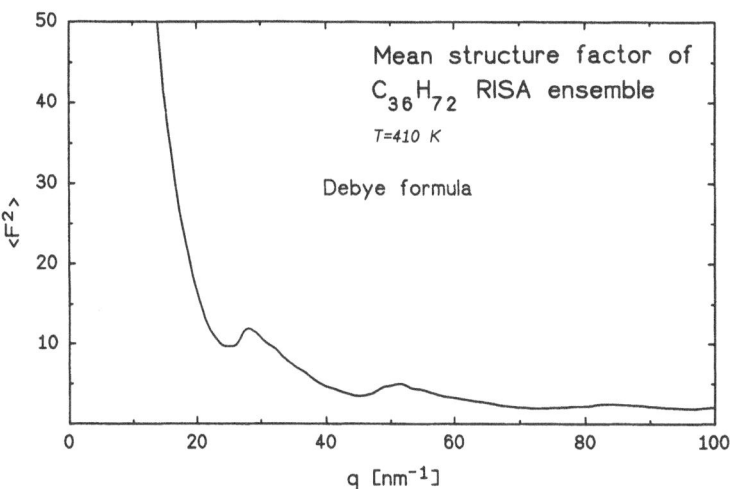

Fig. 3. Mean structure factor of weighted RISA chain segment ensemble, consisting of 500 conformations. The structure factor is normalized to the scattering power of one CH_2 unit

2.6 The Structure Amplitude $\langle F \rangle$ of a Chain Segment Performing Torsional or Rotational Movements

The intermolecular part of the structure factor is determined by the product of $|\langle F \rangle|^2$ and $(S - 1)$. F is the structure amplitude of a chain molecule in system M and depends therefore only on the chain conformation. The symbol $\langle \ldots \rangle$

denotes 'averaging of all chain conformations and orientations' – procedures (1) and (2) – while $\langle \ldots \rangle_s$ performs the isotropic average.

We consequently take advantage of intrinsic symmetries when we now relate rotational or torsional "distortions" to the symmetry axis of chain segments. Such motions are likely to occur without affecting, on average, the global domain structure. It might be well that rotation of the whole segment occurs operating as a "rigid rod" without showing any change of the intrasegmental conformation.

The molecular structure amplitude of a given molecular conformation is [11]

$$F(R, Z, \Psi) = \sum_{n=-\infty}^{+\infty} F_n(R, Z) \exp[in\Psi] \tag{21}$$

with

$$F_n(R, Z) = \sum_j f_j J_n(r_j R) \exp\left[in\left(\frac{\pi}{2} - \Psi_j\right)\right] \exp(iz_j Z) \tag{22}$$

and R, Z, Ψ cylindrical coordinates of structure amplitude
   f_j atomic scattering factor
   r_j, z_j, Ψ_j cylindrical coordinates of atom j
   J_n Bessel function of order n.

Strictly speaking, it is supposed to describe every rotation as a localized motion which should be controlled by interactions between atomic units, molecular groups or the whole segment themselves. These system properties are then phenomenologically represented by the appropriate set of Fourier coefficients. Because we are dealing with WAXS patterns, we observe "quasi-static" configurations. Hence, individual angular motions characterized by individual angular distributions $g_j(\varphi)$ may be the consequence of rotational vibrations as well as of diffusive motions. Both may be differentiated because of the necessity to characterize every stable rotator by defining the "equilibrium values" Ψ_j. The angular distributions $g_j(\varphi)$ are expanded into discrete angular Fourier series, yielding

$$a_{nj} = \int_{-\pi}^{+\pi} g_i(\varphi) \exp(-in\varphi) d\varphi. \tag{23}$$

a_{nj} is the expansion coefficient of order n referring to atom j.

To represent rotation in the most general case we use the Fourier transform

$$\langle F \rangle_g = \sum_{n=-\infty}^{+\infty} F_n'(R, Z) \exp(in\Psi) \tag{24}$$

with

$$F_n'(R, Z) = \sum_j a_{nj} f_j J_n(r_j R) \exp\left[in\left(\frac{\pi}{2} - \Psi_j\right)\right] \exp(iz_j Z). \tag{25}$$

Without any torsion, the g_j-functions are δ-functions; the expansion coefficients a_{nj} are equal to 1. Equation (25) transforms into Eq. (22).

If, on the other hand, free rotation occurs, all a_{nj} are zero except for $n = 0$ where $a_{nj} \equiv 1$. So Eq. (24) reduces to

$$\langle F \rangle_{\text{full rot.}} = F'_0(R, Z) = \sum_j f_j J_0(r_j R) \exp(iz_j Z). \tag{26}$$

This corresponds de facto to a "rotational gas" with the maximum of entropy, because in this idealized model it is supposed that rotation covers isoenergetic angular positions with equal probability.

As we can see from Eq. (26), the main effect coming from chain segment torsions and rotations is that higher orders of the Bessel functions J_n drop out. It depends, however, essentially on the symmetry of the molecule how this effect influences F. We will discuss this important point in more detail in Sect. 4.2.1 'High Temperature Modification of PTFE (Phase I)'.

The basic assumption of our treatment is that polymer structures are topologically characterized by rod-like segments that are packed together to form representative domains. The concept allows us to generate structures of very different shapes with the aid of a unique formalism. The utility of this approach will be discussed with representative examples.

We will be confronted with the question how convincingly it is demonstrated that chain segments are topological subunits which determine and classify the overall features of polymer structures in general.

3 Construction of Two-Dimensional Short Range Order Domains and Their Statistical Properties

Making use of the concept of the effective segment exclusion volume we have now to generate a representative configuration ensemble ('monodomain') of chain segments by arranging mutually interpenetrating cylinders into a two-dimensional distorted cluster. The thickness of the cylinders fluctuates depending on the conformation of the segments involved.

As a first approximation we postulate the existence of an equally sized hard core volume located in the center of each cylinder that defines the excluded volume per segment, according to the degree of interpenetration of neighboring cylinders. This simple construction includes complicated local intersegmental configurations, but it imposes a unique local orientation correlation of segments within the domains of a PE melt. The shape of those impenetrable correlation cylinders is assumed also to be cylindrical.

First we discuss and construct monodisperse two-dimensional arrangements of impenetrable cylinders in terms of radial distance correlation functions, the lateral packing fraction and number density. In the second step, these hard cylinders are covered by the mean electronic density functions of the RISA chain segment ensemble. Last of all, the Fourier transformation and final averaging is performed yielding to the intermolecular part of the monodomain structure factor.

3.1 Construction Schemes

The radial distance distribution in simple atomic and molecular fluids is determined essentially by the exclusion volume of the particles. Zernike and Prins [12] have used this fact to construct a one-dimensional fluid model and calculated its radial distance correlation function and its scattering function. The only interaction between the particles is given by their exclusion volume (which is, of course, an exclusion length in the one-dimensional case) making the particles impenetrable. The statistical properties of these one-dimensional fluids are completely determined by their free volume fraction which facilitates the configurational fluctuations.

3.2 Two-Dimensional Coordination Statistics

It is an unusual problem to construct a homogeneously distorted two-dimensional domain structure which allows us to cover the range of the hexagonal closely packed "crystal" up to a two-dimensional gas. For example, the concept of the paracrystal is based on the philosophy that every state of condensed matter has at least a micro-paracrystalline arrangement of segments within a distorted lattice of arbitrary symmetry with a defined coordination number.

In contrast to that model, we generated statistical homogeneous defect structures with a "broken" coordination number of next neighbors. The exclusion volume of the segments should be accounted for. To our knowledge, there is no mathematical method that allows one to describe the radial distance distribution of such structures analytically. It must be calculated on a computer by generating the structure steadily.

3.3 Algorithm for Generating Homogeneously Distorted Two-Dimensional Monodomains by Introducing Three-Point Coordination Statistics

Any homogeneously distorted two-dimensional coordination scheme should be based upon the distance correlation statistics between next neighbors at least. In the case of a two-dimensional lattice construction, this distance correlation principle has been used by Hosemann and coworkers [13] to generate micro-paracrystals of finite size with the help of a computer. The construction procedure (known as 'spiral-paracrystal') terminates if a coordination point cannot be assigned to a lattice point.

The construction scheme we present in our work does not require a lattice at all. Due to the fact that the distance correlation between three coordination points is considered, the algorithm generates new coordination points in a unique manner. Any correlation of these coordination points to a lattice is disregarded as the generation procedure runs continuously.

In the following we describe the coordination generation procedure of a monodisperse system of hard discs in detail with an example run.

1. First of all, a distance correlation function $g_{NN}(r)$ of next neighbors must be given. This function has to meet the following conditions:

$g_{NN}(r) = 0$ for $r \leq r_H$ (disc diameter)
$g_{NN}(r)$ maximum   for $r = \langle r \rangle$ (mean distance between next neighbors)
$g_{NN}(r) \to \infty$ for $r \to \infty$ (no correlation at large distances)

For our purposes we used a Gaussian distribution shifted in the ordinate direction towards lower values (Fig. 4). The standard deviation σ of this distribution is correlated somehow with the extent of distance fluctuations in the system.

2. Defining a starting point, two discs are placed with distance $\langle r \rangle$ to each other (Fig. 5a).
3. Two distances r_1 and r_2 are generated obeying the distance distribution given by $g_{NN}(r)$.
4. The new coordination points are defined by the intersection of the two circles with radius r_1 and r_2, respectively (Fig. 5b).
5. A disc is inserted at the new coordination point. If there is an overlap with other discs, the new position is rejected completely, otherwise it is accepted.
6. To continue, a new pair of discs is chosen serving as a base for the construction of further coordination points (Fig. 5c, base pair 1–3 for example).

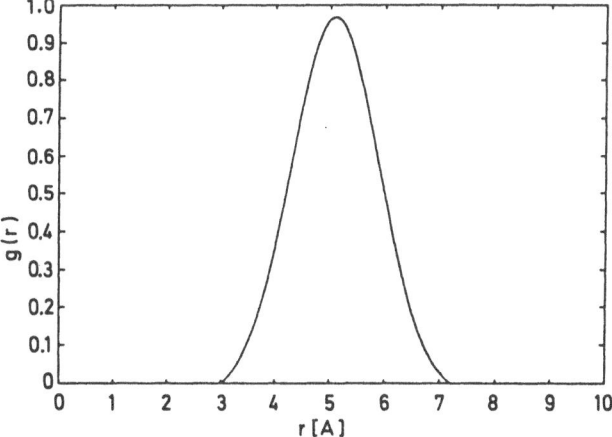

Fig. 4. Distance distribution of next neighbors, approximated by a shifted Gaussian distribution (see text). $r_{hard} = 3$ Å, $\langle r \rangle = 5.1$ Å, $\sigma = 0.8$ Å

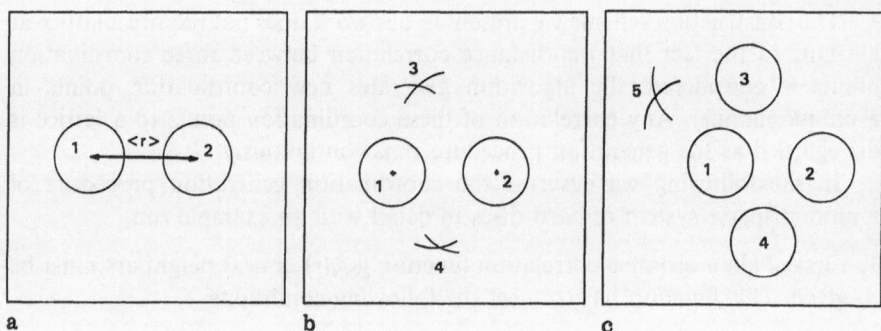

Fig. 5a. Initial setting of two discs at distance $\langle r \rangle$. **b.** Determination of new disc coordination points by the method of circle intersection. **c.** Transition to a new base pair

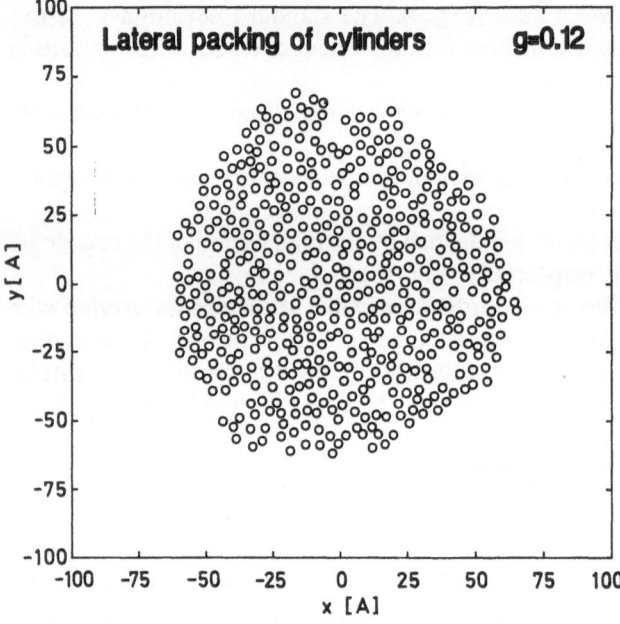

Fig. 6. Example of a two-dimensional disc arrangement according to the construction scheme

7. Finally, steps 3 to 6 are repeated until a given system size is reached (Fig. 6). The construction terminates only for technical reasons (CPU time), not for physical reasons.

The whole construction procedure resembles a non-lattice-like self-avoiding random walk used in many MC- and MD-simulations.

To obtain the radial distance correlation function one has to take the average of many system configurations. The fluctuation parameter g can be derived from these functions easily.

As a consequence of exclusion, the coordination number shows very peculiar behavior (Fig. 7). For example, six percent relative quadratic fluctuation yields to a bimodal distribution (Fig. 8). This indicates that every statistical homogeneous coordination needs larger distance fluctuations of segments lying in the

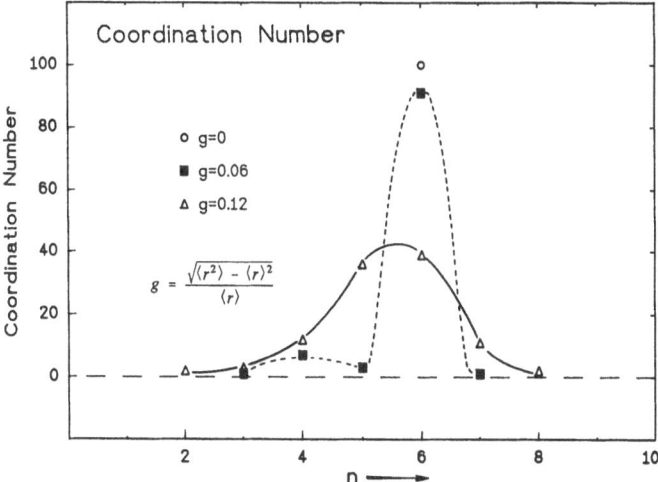

Fig. 7. Distribution of the coordination number, shown for different fluctuation parameters g. For g = 0.06, a bimodal distribution results (see Fig. 8)

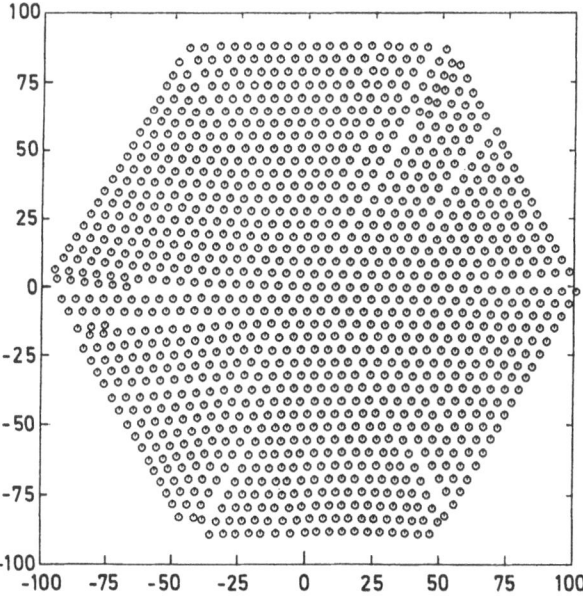

Fig. 8. 'Symmetry break' of the hexagonal structure

range of g 0.1 to 0.2. Increased fluctuation is accompanied by an increased relative mean free volume. This example indicates the role of the exclusion volume on phase transitions being in principle analogous to the van der Waals-equation of real gases.

The most crucial condition the computer construction must be submitted to, is the adjustment of the packing fraction[1] to the mean density of real structures. This physical condition gives the model a relevance which exceeds its purely geometrical origins.

This construction procedure does not allow us to derive the mean packing fraction from the model parameters r_H, $\langle r \rangle$ and σ easily. It should be possible, however, to give an empirical correlation between those parameters and the packing fraction.

For technical reasons, it is simpler to use Monte-Carlo methods with a constant packing fraction.

3.4 Two-Dimensional Liquid-Like Distance Statistics of a Given Packing Fraction Generated by Monte-Carlo Techniques

The fit of the model packing fraction to the macroscopic density is the essential point of our model, as already mentioned. That is why we chose a Monte-Carlo method to obtain two-dimensional liquid-like distance statistics of hard discs. The procedure we used is exactly the same as used by Metropolis et al. [14] with the addition of averaging a large number of system configurations.

We started from hexagonal configurations of 19 by 22 discs with lattice constant a and number density

$$n_0 = \frac{2}{a^2 \sqrt{3}} \qquad (27)$$

(see Fig. 9) and introduced periodic boundary conditions.

The thermalization of the system follows the well-known Metropolis rules (importance sampling procedure) [14] obeying the hard core overlap, but no additional interactions. The degree of thermalization can be controlled by observing the melting factor distribution. The melting factor is a means of judging if the model system has already reached thermodynamic equilibrium. For our purposes we defined

$$M = \frac{1}{2} \sum_{i=1}^{N} \left[\cos \frac{2\pi m_x}{L_x} 2x_i + \cos \frac{2\pi m_y}{L_y} y_i \right] \qquad (28)$$

as melting factor. L_x and L_y are the lengths of the base cell, m_x and m_y the number of lattice points in x- and y-direction, x_i and y_i the lattice point

---

[1] The packing fraction is given by the ratio (exclusion volume/total volume).

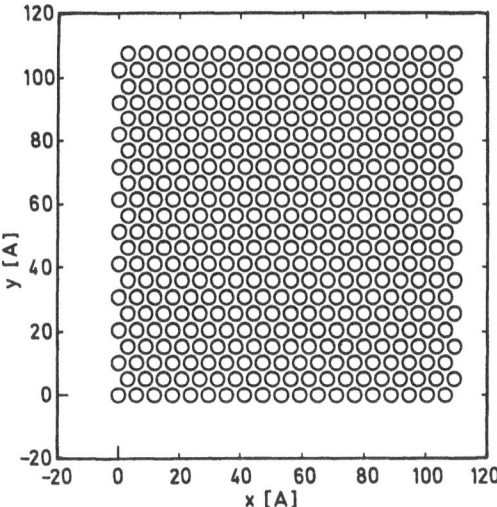

Fig. 9. Initial configuration of all MC simulation runs with $19 \times 22 = 418$ discs on a hexagonal lattice

coordinates. When the system reaches thermodynamic equilibrium and shows liquid-like properties, the melting factor distribution as an average of many configurations approaches a Gaussian distribution with standard deviation $\sqrt{N}$, centered around 0 (Fig. 10).

After the thermalization run, configuration sampling begins to obtain RDCF and coordination number distribution. Figure 11 shows one single system configuration, mean RDCF, coordination number distribution and configuration structure factor S(q) with an example.

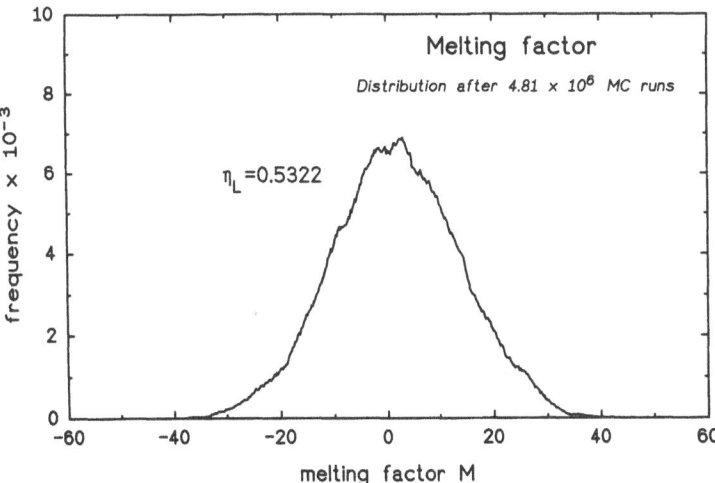

Fig. 10. Distribution of the melting factor after the thermalization run

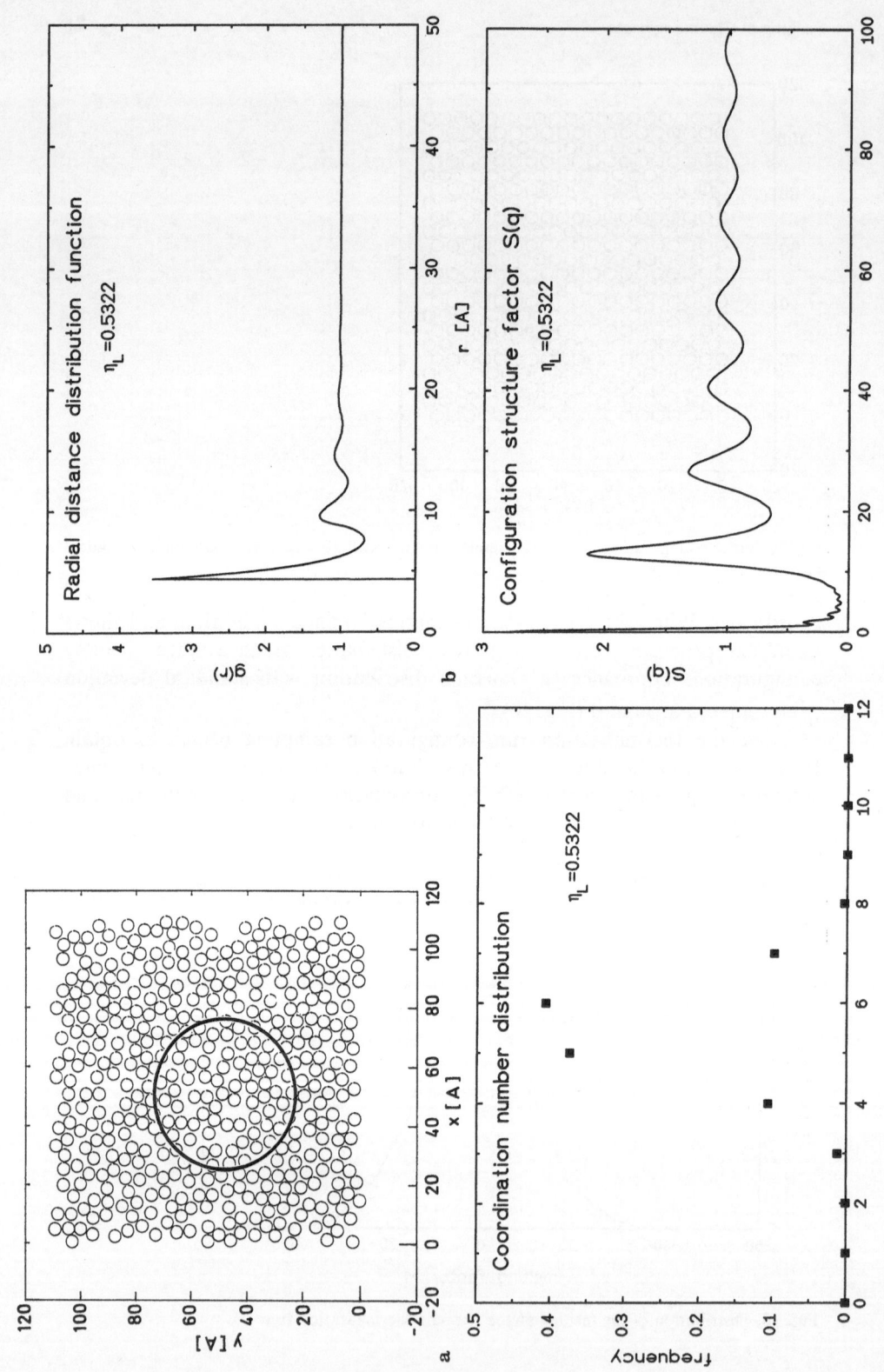

Radial distance distribution function
$\eta_L = 0.5322$
r [A]
g(r)
b

Configuration structure factor S(q)
$\eta_L = 0.5322$
S(q)

Coordination number distribution
$\eta_L = 0.5322$
x [A]
y [A]
frequency
a

4 Results and Discussion

4.1 PE-Melt

4.1.1 Experimental Setup

It is a well-known fact that precise determination of radial density correlation functions from WAXS-patterns with good resolution requires the facility to measure up to high values of the scattering vector. To achieve this, we used Zr-filtered Mo-K$\bar{\alpha}$-radiation. The experimental data were obtained with a conventional texture goniometer (PW 1078, Philips) which has been modified to work in transmission geometry. The sample, a commercial linear polyethylene (Lupolen L6041D, BASF, $\bar{M}_w \approx 2 \times 10^5$ g mol^{-1}, density 0.960 g/cm^3 at room temperature, thickness 0.65 mm) was mounted on a special heating device which can be heated up to approx. 180 °C. The sample geometry is shown in Fig. 12. It should be mentioned that the sample has been tilted at $\Psi = 30°$ to the incident X-ray beam. This became necessary due to the special geometry of the sample holder.

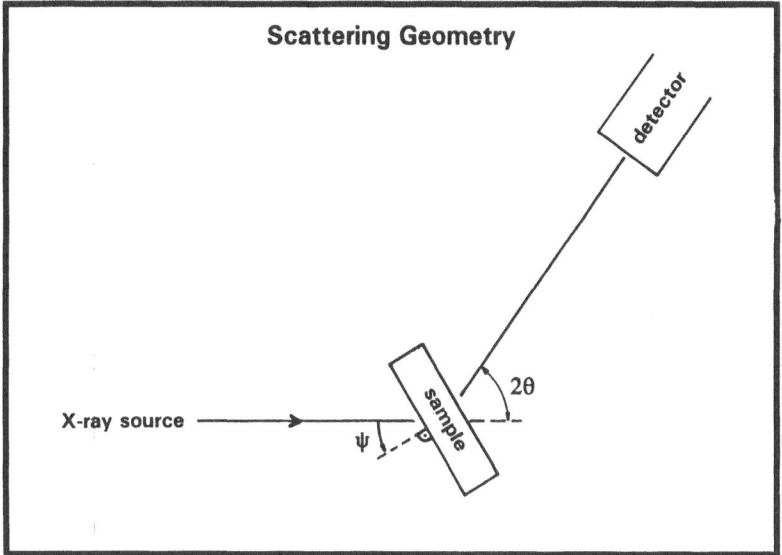

Fig. 12. Schematic view of the scattering geometry. The unusual tilt of the sample became necessary due to the special geometry of the sample holder

Fig. 11. a. Disc arrangement of one ensemble configuration; **b.** RDDF, averaged over 4×10^6 configurations; **c.** Coordination distribution of entire ensemble; **d.** Mean structure factor

The temperature was kept constant at 153 °C and atmospheric pressure during the measurement. The detection system consisted of a Xenon-filled proportional counter in combination with pulse height discrimination and dead time correction (PW 1710, Philips). Pulse preset mode with 2000 counts minimum was selected to reduce statistical errors. After correcting for air scattering, the relative error was smaller than 5% in the worst case. The 2θ-region extended from 2° to 67° yielding a scattering vector region from approx. 4 to 95 nm^{-1}.

The experimental data were corrected for air scattering, polarization, absorption and Compton scattering applying well-known correction procedures (for example [15], [16]). Then the data were transformed into q-space using

$$q = 2\pi \frac{2\sin\theta}{\lambda}$$

and normalized to electron units applying the normalization procedure introduced by Krogh-Moe [17]. The corrected and normalized scattering pattern is shown in Fig. 13. From this we calculated the electronic radial density difference function (RDDF)

$$G(r) = 4\pi r^2 [\rho(r) - \rho_0] = \frac{2r}{\pi} \int_0^\infty qi(q)\sin(rq)dq \qquad (30)$$

in which i(q) is the reduced scattering function given by

$$i(q) = I_{corr}(q) - \sum_k f_k^2 \qquad (31)$$

We chose the electronic RDDF because in our model the electronic density of the structure elements is cylindrically symmetrical. However, the calculation of

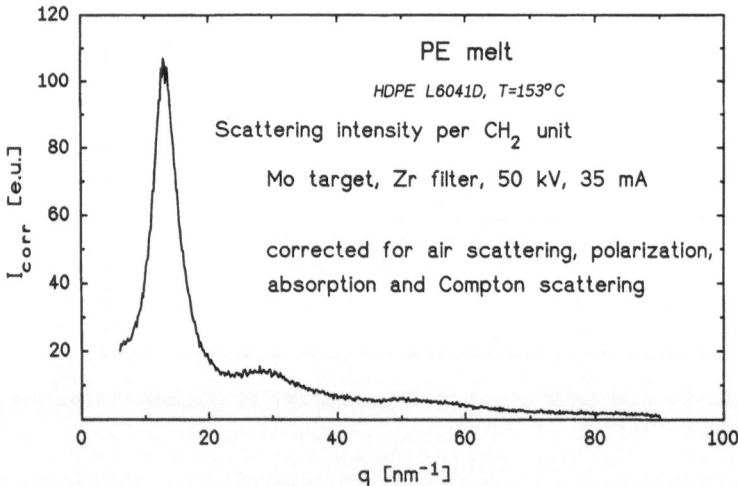

Fig. 13. Measured scattering intensity of HDPE melt. Intensity corrections for air scattering, polarization, absorption, sample geometry and Compton scattering have been performed

atomic RDDF's is sensible only if the electronic density of the structure elements has spherical symmetry (this is, for example, approximately true for CH-units [18]). To minimize termination errors we used a sampling procedure given by Lovell, Mitchell, and Windle [19] and the damping factor introduced by Lorch [20]. The resulting electronic RDDF is shown in Fig. 14.

The intramolecular distances belonging to fixed C–C-distances along the chain (1.54 Å and 2.54 Å) cannot be resolved due to the smearing effect of the electronic density. The peaks coming from intermolecular distance correlations can be seen clearly, the range of the appropriate short range order is about 25 Å which is comparable with results from other authors [3, 7].

4.1.2 Mean Electronic Density of PE Segments and Determination of Characteristic Segment Length

The essential step to simulate a realistic short range order region with our model is to occupy the local segment exclusion volume with the mean electronic density of the chain segments yielding the correlation cylinders introduced in Sect. 2.

To do this, we generated CH_2 sequences of appropriate length using RISA. Then we fitted short CH_2 sequences (consisting of 5–10 CH_2 units) out of the segment ensemble[2] into the correlation cylinders taking the number of CH_2 units per cylinder as fitting parameter. The fits were performed by defining an

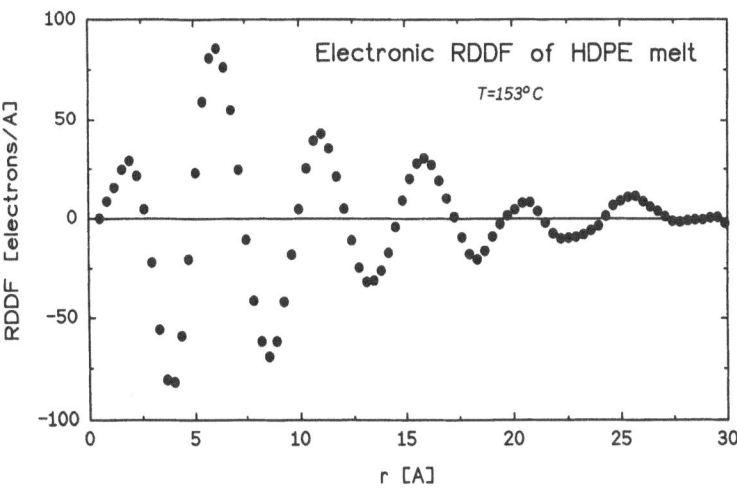

Fig. 14. Experimental RDDF from HDPE melt at T = 153 °C. The finite resolution in r-space results from the finite q-range

---

[2] This procedure avoids boundary effects.

appropriate segment end-to-end vector (Fig. 15a) and rotating and shifting this 'chord' into the cylinder (Fig. 15b), thus minimizing the 'electronic radius of gyration'

$$R_e^2 = \frac{\sum_k r_k^2 Z_k}{\sum_k Z_k} \tag{32}$$

with r_k: perpendicular distance projection of atom k onto cylinder axis
Z_k: order number of atom k.

The mean electronic densities obtained by this procedure are shown in Fig. 16 for segment lengths of 5 to 10 CH_2 groups.

The electronic density in the radial direction (perpendicular to the cylinder axis) gets delocalized more and more with increasing segment length because there are more CH_2 units in the gauche position included enlarging the cylinder diameter.

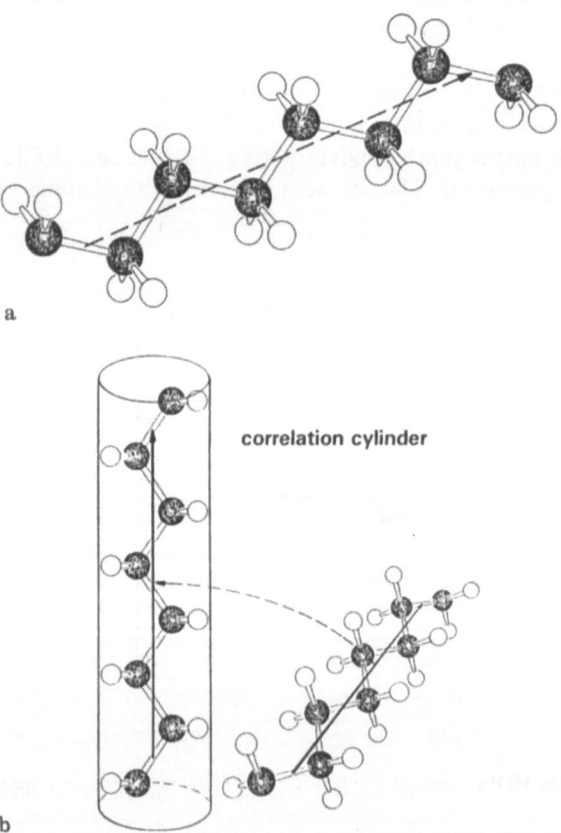

Fig. 15a. Definition of segmental end-to-end-vector (see text for details). **b.** Fit into correlation cylinder

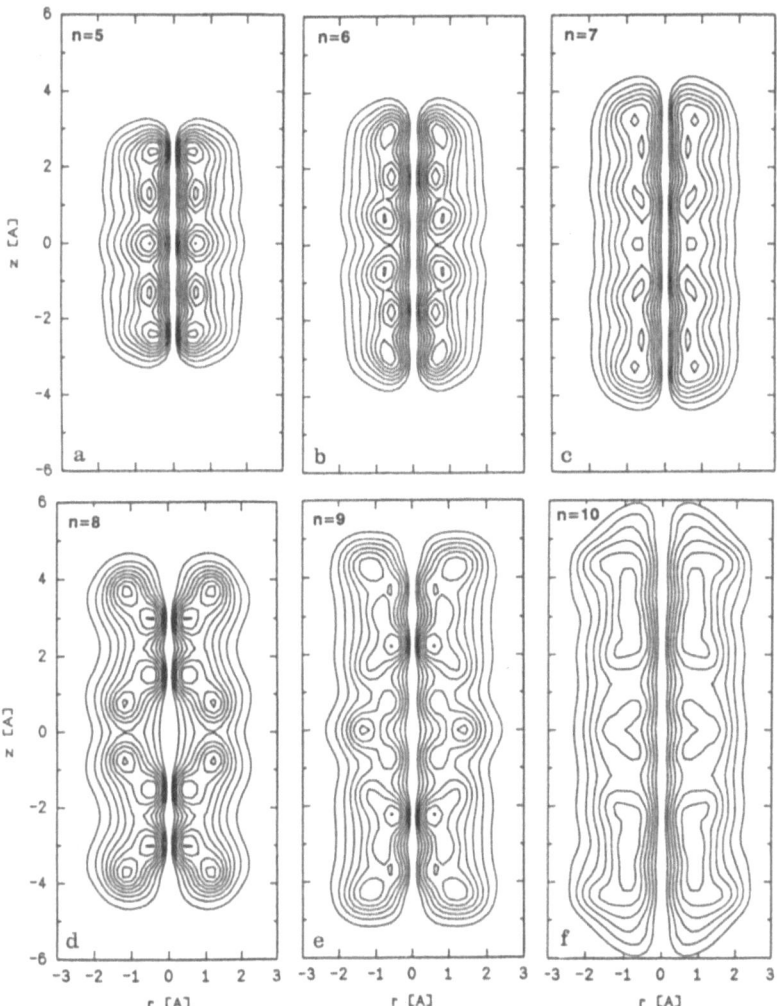

Fig. 16. Cylindrically symmetric electron densities for several numbers of CH_2 units within the correlation cylinder

From the mean electronic density $\langle \rho(r, z) \rangle$ the cylinder symmetric structure amplitude $\langle F(q_r, q_z) \rangle$ can be easily obtained by the Fourier-Bessel-transformation:

$$\langle F(q_r, q_z) \rangle = \int_{-\infty}^{+\infty} \int_{0}^{\infty} \langle \rho(r, z) \rangle J_0(rq_r)\cos(zq_z)2\pi r \, dr \, dz \qquad (33)$$

The square of the structure amplitude in radial direction is shown in Fig. 17 for segment lengths 5 and 10. Even for short segment lengths, $\langle F(q_r) \rangle$ drops rapidly against zero due to the conformational and rotational average within the

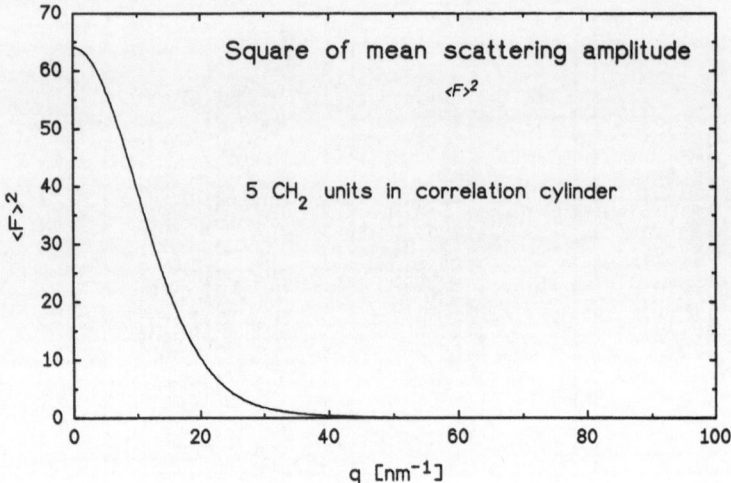

Fig. 17. Square of the mean segment structure amplitude in the lateral direction for 5 CH$_2$ units within the correlation cylinder. The structure amplitude drops off very quickly due to the strong electron delocalization

cylinder. It is exactly this function which allows us to determine the characteristic segment length being responsible for the ratio of intra- to intermolecular scattering contribution in the WAXS pattern.

4.1.3 Calculation of the Structure Factor and Fit to Our Model

The calculation of the model structure factor in Eq. (19) requires the knowledge of three terms: First, the average molecular structure factor $\langle |F| \rangle^2$, second, the square of the average molecular structure amplitude $|\langle F \rangle|^2$ and last, the configuration structure factor S.

In order to fit our structure model to the experimental data, we have to vary the model parameters packing density and cylinder length and examine the effects on the RDDF of our model via the structure factors S and F. One can easily see that an increase of packing density causes the intermolecular distance correlation range to increase because mobility and free volume are reduced. In other words one can calculate a realistic value of packing density (i.e. cylinders per volume) from the experimental RDDF (Fig. 18).

On the other hand, a variation of the equivalent structure unit length (cylinder length) causes the intermolecular peaks belonging to next neighbors and next-but-one neighbors to change their positions slightly and their intensity values dramatically without disturbing essentially the lateral short range order (Fig. 19).

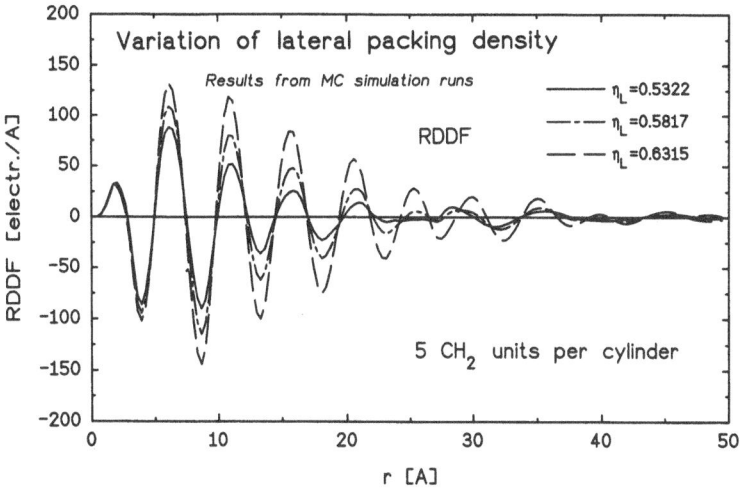

Fig. 18. RDDF of the model with variation of the packing density parameter η_L. The range of distance correlation increases with increasing η_L

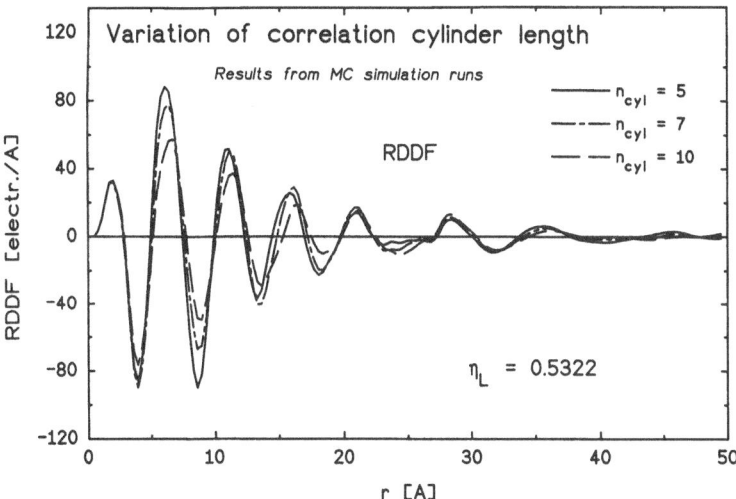

Fig. 19. Variation of the cylinder length n_{zyl} (number of CH_2 units). The number of next neighbor distances decreases with increasing cylinder length

The best fit to the experimental RDDF yields

$$\eta_L = 0.5322$$

$$n_{cyl} = 5$$

(34)

where n_{cyl} is the number of CH_2 groups within the equivalent structure unit. Both the experimental and the model RDDF are plotted in Fig. 20. During the

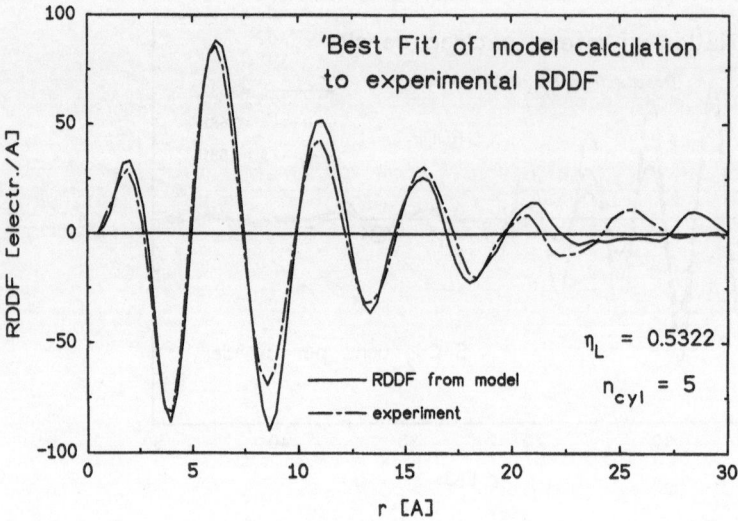

Fig. 20. Model-RDDF compared with experimental RDDF

fit procedure we took care that the intensity ratio of the intramolecular peak at approx. 2 Å (combination of two peaks at 1.54 Å and 2.54 Å respectively – see above) and the first intermolecular distance correlation at approx. 6 Å corresponded to that of the experimental RDDF.

The structure factor of the model short range order regions and the experimental WAXS pattern are shown in Fig. 21. The strong Guinier scattering at

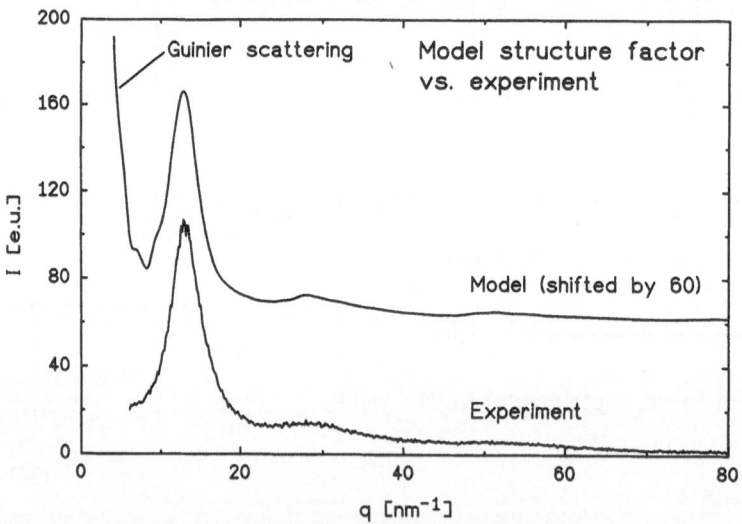

Fig. 21. Comparison of model structure factor and experiment. The strong scattering contribution at small q-values comes from the finite size of the model structure

low q-values results from the finite size of our model. Thus, we considered several Guinier correction terms but did not obtain satisfactory results.

It is an essential test of our model that the number density of CH_2 groups corresponds to the macroscopic density. To verify this, we calculate the experimental CH_2 density from the macroscopic density using

$$n_{CH_2}^{exp} = \frac{\rho_0 \cdot N_A}{14 \text{ g/mol}} = \frac{0.777 \text{ g/cm}^3 \cdot N_A}{14 \text{ g/mol}} = 0.0334 \text{ Å}^{-3} \tag{35}$$

where the density was obtained from values of the specific volume of PE given by Stewart and von Frankenberg [21].

The CH_2 density of our model can be easily calculated from the interchain packing density and the mean intrachain CH_2 spacing. The former was obtained from the fit of the RDDF's, the latter can be extracted from the mean electronic density (Fig. 16). If we consider complete space filling of the model short range order regions, we finally get

$$n_{CH_2}^{mod} = n_{packing} \cdot n_{chain} = 0.033 \text{ Å}^{-2} \cdot 0.833 \text{ Å}^{-1} = 0.0275 \text{ Å}^{-3} \tag{36}$$

which is only 82% of the experimental value.

Until now, we have assumed that the microstructure of the melt can be modelled by short range order regions consisting of nearly parallel chain segments which are about 6 Å long. However, the MC calculations clearly show that enough free volume exists within the domains which can be occupied by additional chain segments. The free volume can be filled with chain segments which do not contribute to any constructive interference like a low-concentrated 'defect gas' consisting of entanglements, chain ends, non-parallel chain segments pairs or chain folds, for example. The consequences of these gas-like defects cannot be seen in the RDDF which is only sensitive to deviations from the mean electronic density.

Another reason for the low density comes from the simplification that we have considered only monodisperse packing models until now. The introduction of fluctuating chain segment exclusion volumes leading to polydisperse packing models would be more realistic, of course. The distance correlation range of polydisperse packing models is smaller than in monodisperse systems with the same packing fraction, thus allowing the packing density to increase in the polydisperse case.

To summarize this section, we conclude from our simulations and measurements that the short range order in a PE melt is mainly characterized by segment-to-segment orientation correlation of next neighbor segments of length 6 Å with liquid-like radial distance statistics. The range of radial distance correlations is about 25 Å. If the model assumptions are taken seriously, the short range order domains must include chain defects (entanglements, chain ends and folds) as well as orientation correlation defects (non-parallel segment pairs). Obviously it is not possible to reflect the effective segment exclusion volume onto a monodisperse arrangement of correlation cylinders. The high value of the melt density together with the results of WAXS measurements demand the discussion of polydispersity.

4.2 Segment Conformation and Dynamics in Polymer Mesophases

In the case of the PE melt, the largest part of the configuration entropy is included in the variety of chain conformations. In our model of the PE melt, these chain conformations cause the diameter of the correlation cylinders in which the chain segments are embedded to fluctuate. This leads necessarily to distance fluctuations of next neighbor segments because of steric hindrance of the local segment exclusion volume. Therefore, the inter- and intramolecular distance correlations cannot be discussed independently.

The correct calculation of WAXS patterns of 'Condis phases' and smectic LC phases would provide us additional information, because in some of these systems, intramolecular torsions and rotations occur which do not affect the lateral chain packing. We chose the high-temperature phase of PTFE (phase I) and a S_E-phase of a LC main-chain polyester as model systems.

4.2.1 High Temperature Modification of PTFE (Phase I)

The transition of PTFE (poly-tetrafluorethylene) from phase II to phase I in the range of 19 °C to 30 °C has been investigated extensively since the work of Bunn and Howells [22–29]. Below 19 °C (phase II) the PTFE structure cannot be assigned to a point group [30]. The molecular conformation in this modification is a 13_6-helix [30–32]. Increasing temperature causes the chain segments to perform torsional and rotational motions around the helix axis. Above 30 °C the PTFE structure is transformed into the disordered phase I with a crystal symmetry close to hexagonal [30]. The helix conformation modifies slightly to 15_7. The dynamics of the segment motion have been determined by several methods such as NMR absorption [26], neutron scattering [33] and Raman scattering [34].

The effects of this transition on the WAXS pattern have been investigated in detail by Clark and Muus [27, 28]. The local rotation of chain segments around the helix axis smears out the hkl-reflections along the layer lines leaving only the equatorial hk0-reflections and the intramolecular helix reflections near the meridian (Fig. 22).

The molecular mechanism of local segment rotations can be explained by the occurrence of twin reversals [33, 34] which are induced thermally due to the unusual course of the conformational potential [35]. Those twin reversals are torsion defects causing the helix conformation to change from left-handed to right-handed and vice versa. They are built into the PTFE helix without changing the direction of the molecular long axis (Fig. 23). Additionally, the long

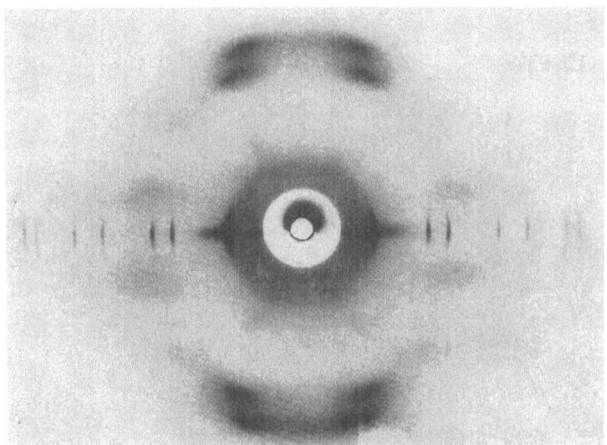

Fig. 22. WAXS fiber pattern of PTFE phase I at T = 30 °C [27]

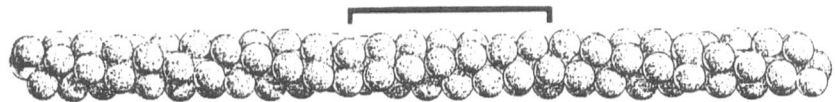

Fig. 23. Twin reversal built into a 15_7 PTFE helix

range dipole forces caused by the C–F dipole moment support an elongated molecular shape and high chain stiffness.

To calculate the WAXS fiber pattern of PTFE phase I, we assumed that the CF_2 units are equally distributed around the chain axis due to the high mobility of twin reversals in the high-temperature phase. In other words, the electron density of the PTFE chains is cylindrically symmetric in space average. Thus we are able to calculate the mean molecular structure amplitude $\langle F \rangle$ according to Eq. (26).

The WAXS pattern can be calculated in a quantitative way when we arrange 10×12 PTFE stems of length 40 Å (two monomers) on a two-dimensional hexagonal lattice with a = 5.66 Å (Fig. 24). Therefore, the lateral size of these 'Condis-crystals' is at least 60 Å.

The essential result of the model calculation is that in phase I the molecular electron density is delocalized cylindrically symmetric in space and time average. This is achieved by a complete revolution of every CF_2 unit around the chain axis within an axial range of 40 Å.

From this, one could assume that rotational space averaging always causes hkl-reflections to disappear and hk0-reflections to remain unchanged. However, as we will show in the next section with another model system, this is highly

Calculated WAXS-pattern of PTFE (Phase I)

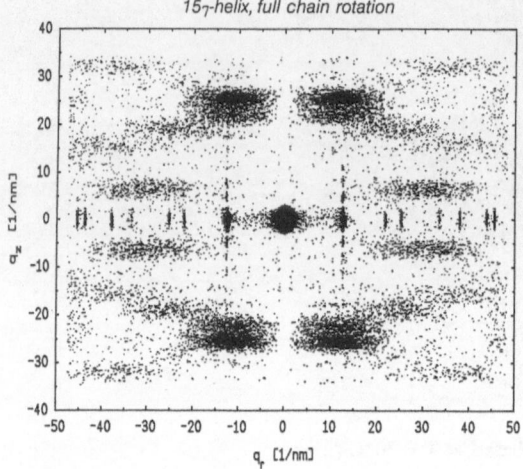

Fig. 24. WAXS fiber pattern of a hexagonal chain arrangement consisting of 10×12 PTFE stems of length 40 Å, assuming perfect orientation. The sharp spots are the consequence of finite model size

dependent on the molecular symmetry. In the case of the 15_7 helix conformation, Bessel functions of orders 0, 15, 30 . . . contribute to the molecular structure amplitude in the equatorial direction, if no chain rotation occurs at all. If the chains perform a complete rotation, orders 15, 30 . . . of the Bessel function drop out and only order 0 remains. However, the contribution of the higher orders to the structure amplitude is neglectible, order 0 dominates.

The example presented also underlines the importance of rotatory segmental motions indicating a certain autonomy of intramolecular processes. It is, of course, also the increased entropy that is behind the outstanding thermal stability of PTFE.

4.2.2 Smectic Phase of an LC Main Chain Polyester

With our model we are in a good position to discuss the LC-main chain polymers in a smectic phase. It must be seen as a typical feature that only one single very strong interference appears. A representative WAXS-pattern of a smectic LC-polyester (Fig. 25, Fig. 26) is shown in Fig. 27.

According to symmetry conditions it is reasonable to consider rotation of distinct units or of a whole mesogenic group. It is clear from the sketch in Fig. 28 that the r_{kl}'s will, in this way, be changed substantially. However, free motions of this type are only possible if the lateral distance of the mesogenic units is large enough. Hence, we have to consider the correct packing of the mesogenic units which are, in the present example, built into a two-dimensional lattice of orthorhombic symmetry. These are good reasons for assuming that the lattice is

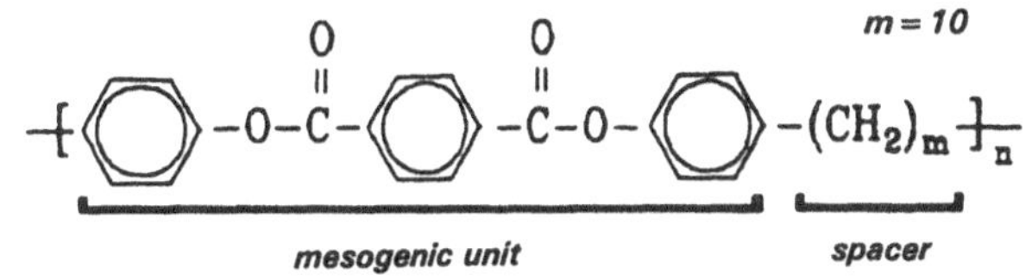

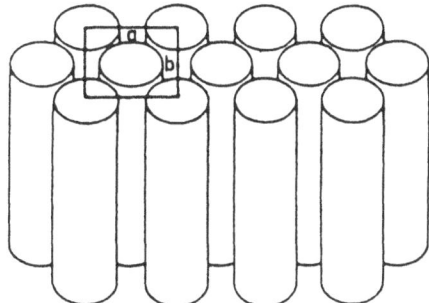

Fig. 25. Chemical structure of the LC main chain polyester

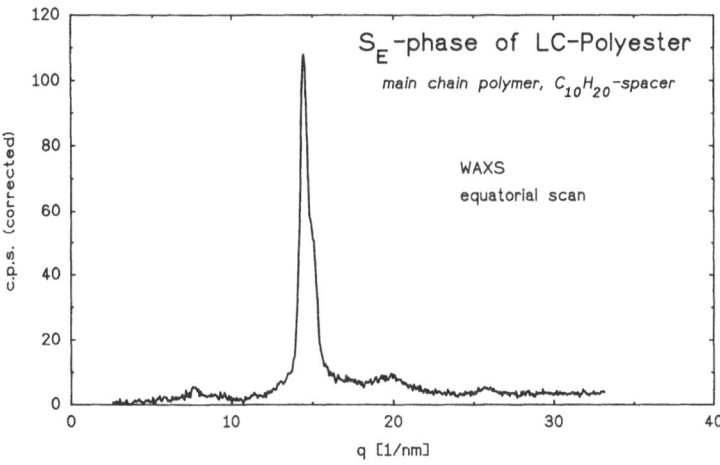

Fig. 26. Arrangement of the mesogenic units in the S_E phase

Fig. 27. Equatorial WAXS intensity of S_E-phase of a uniaxially oriented LC main chain polyester. 110- and 200-reflections contribute to the main reflection. No second order (220 and 400) appears

not distorted so much. Small lateral collective motions would produce Debye-Waller-effects that cannot be observed under any circumstances because of the predominance of rotational motions.

Taking as a state of reference the extended mesogenic units shown in Fig. 28, we consider rotation of benzenes in respect to the symmetry axis as indicated in

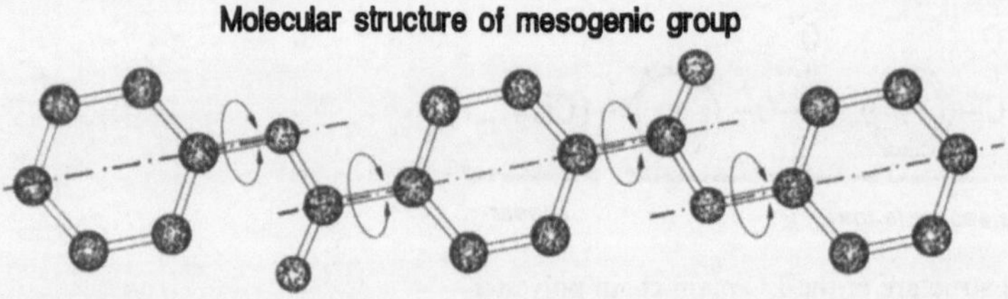

Molecular structure of mesogenic group

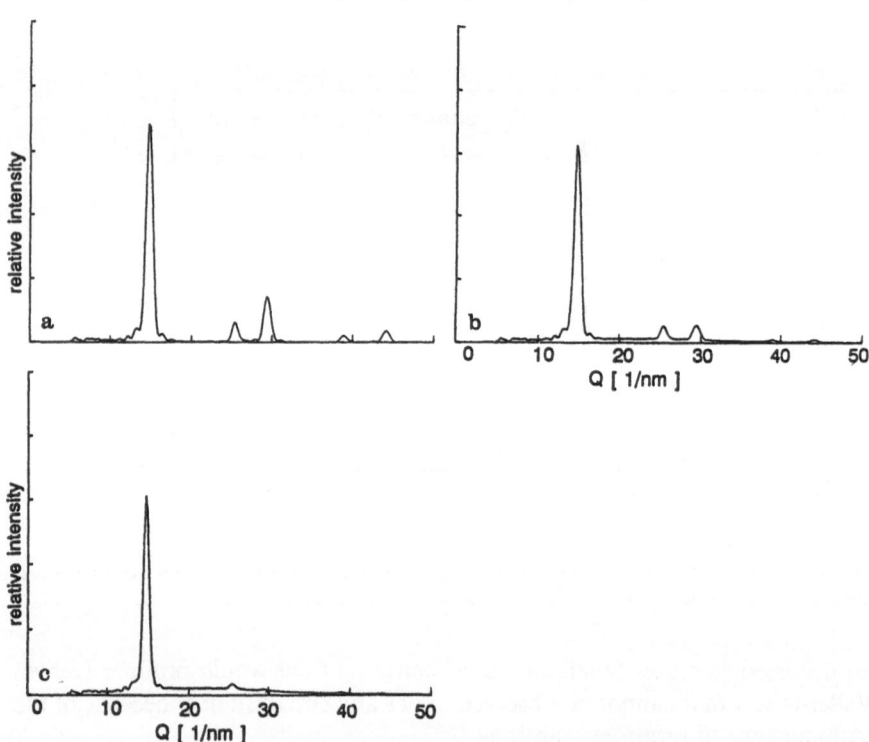

Fig. 28. Schematic representation of rotational degrees of freedom in the mesogenic unit. Only C- and O-atoms are drawn for reasons of clarity

the drawing. In a diffusive model each of the "rotational isomers" produced by independent rotation of the benzenes is believed to be realized with the same probability. In these circumstances, the maximum rotational angle $\Delta\varphi$ is left as the essential fitting parameter. It is now demonstrated with the calculated WAXS-diagrams in Fig. 29 that by increasing $\Delta\varphi$, all the interferences at q's beyond the single "main peak" are increasingly smeared out accompanied with a diffuse very broadly scattered background.

Fig. 29a–c. Calculated WAXS intensity of the S_E phase with different amounts of torsion amplitude of the aromatic rings: **a.** $\Delta\varphi = 0°$; **b.** $\Delta\varphi = 30°$; **c.** full rotation. See Eq. (24)

Accordance with the experiment (Fig. 27) is only achieved when there is diffusive and free rotation of benzenes. The smectic polyester LC-phase is apparently characterized by

(a) layers of extended meosgenic groups showing basically a perfectly ordered "two-dimensional" lattice. The spacer units are squeezed out of these layers for thermodynamic reasons (spacer and mesogenic units are "lattice-incompatible")
(b) the benzenes seem to be capable of occupying all the rotational sites independently allowing, nevertheless, the mesogenes to be packed in a lattice to reproduce the observed angular position of the main interference.

It cannot be identified by means of our analysis how the various rotational isomers come about. We only know that the underlying motions should be largely uncorrelated. Therefore there is also no X-ray coherence between neighboring layers of extended mesogenic units. One could come to the conclusion that free and independent rotation of the benzenes is important because of producing the maximum entropy of gas-like dynamics within the crystallographic layers, thus minimizing the free enthalpy of the smectic phase. These layers may be considered as the basic cooperative units forming the superstructure of the smectic phase. This is also dictated by the necessity of squeezing the spacer-groups which are linked in the chain out of the mesogenic layers. Hence, forming mesogenic membranes alternated with disordered spacer layers reveals a superstructure that is controlled by the chemical structure of the LC main chain polymers. The quality of the superstructure can be characterized by an analysis of the "middle-angle" interferences.

5 Conclusions

We have shown with representative examples that rotation of larger molecular units or of chain segments can be identified by an analysis of WAXS-patterns of liquid crystalline or crystalline polymer systems. Rotation is indicated by a substantial loss of Bragg-interferences in the "high-q-range". The representative examples shown here give outstanding evidence of the importance of rotational modes of motion. These motions are, of course, restricted to segments of finite length. It is an interesting matter to see that smectic LC-main chain phases are likely to be established because of the free or almost unconstrained rotation of subunits. These dynamics should be the reason for the reduced viscosity observed in such systems. Strictly speaking, the smectic phases discussed are crystals with a well established two-dimensional lattice constituted by layers of finite thickness. The rotation, if it is diffusive, displays, to a first approximation, gas-like behavior. This brings the cooperative mean-field interaction of the mesogenic units fully into play with a rotational averaged lattice potential.

Whether the segments in the layers are orthogonal or not depends on the orientation of the axis of rotation with respect to the axis of the whole mesogenic unit or the whole segment and the way the mesogenic units are coupled to the polymer chain.

We conclude that the existence of side chain smectic LC-polymers is basically dependent on the length of the spacer. This length must be sufficiently large to allow rotation of segments in the mesogenic units as well as to allow a perfect separation of these units from the incompatible spacer-chain-complex. This hypothesis is qualitatively justified by finding analogous characteristics of WAXS-patterns as shown in this paper.

The intramolecular chain structure in a polymer melt determines, by its rotational potentials, the size of chain segments which is in any case equal to or larger than that of the free chain. Because of having found by neutron-scattering experiments that the radius of gyration of chains in the melt is close to the one of the free chain, intramolecular correlations should not be dramatically modified with respect to the freely jointed chain (Kuhn segment). For this reason, it is obvious that the approximately correct intramolecular conformation of the X-ray representative segment can be obtained by the use of single chain segment calculations. Despite some discrepancies in the q-range which is supposed to be mainly determined by intramolecular r_{kl}'s, the longitudinal correlation length cannot be very much different to the one in a polymer chain.

Our results suggest that every packing is locally bound a priori to the exclusion of a hardcore volume of each segment. The stability of the melt needs a sufficiently large intermolecular free volume to allow on an average free diffusive rotation of chain segments. These modes of motion are the reason for organizing monodomains as stacks of more or less parallel segments at the minimum expense of conformational freedoms. The longitudinal extension is defined by intramolecular properties, while the size in the lateral direction is controlled by the mean-field interaction of rotationally averaged short rod-like segments. The basic symmetry as represented by packing these anisometric rods together cannot be changed even by very delicate local interchain configurations. The topological feature of all polymer structures turns therefore out in an evident manner to be induced by the existence of representative rod-like segments the cylindrical symmetry of which makes rotational modes of segmental motions a determining fact. This was one of the reasons for developing our model for describing polymer structures in the way shown in this paper.

6 References

1. Stein RS, Hong SD (1976) J Macromol Phys B12: 125
2. Fischer EW, Strobl GR, Dettenmaier M, Stamm M, Steidle N (1979) Faraday Discuss Chem Soc 68: 26
3. Longman GW, Wignall GD, Sheldon RP (1979) Polymer 20: 1063

4. Benoit H (1976) J Macromol Sci Phys B12: 27
5. Wignall GD, Ballard DGH, Schelten J (1976) J Macromol Sci Phys B12: 75
6. Deas HD (1952) Acta Cryst 5: 542
7. Mitchell GR, Lovell R, Windle AH (1982) Polymer 23: 1273
8. Abe A, Jernigan RL, Flory PJ (1966) J Am Chem Soc 88: 631
9. Hägele PC, Pechhold W (1970) Koll Z u Z Polym 241: 977
10. Bautz G, Leute U, Dollhopf W, Hägele PC (1981) Coll Polym Sci 259: 714
11. Vainshtein BK (1966) Diffraction of X-rays by chain molecules, Elsevier, Amsterdam
12. Zernike F, Prins JA (1927) Zeitschr d Phys 41: 184
13. Janke M, Hosemann R (1978) Progr Coll Polym Sci 64: 226
14. Metropolis N, Rosenbluth AW, Rosenbluth MN, Teller AH, Teller E (1953) J Chem Phys 21: 1087
15. Schubach HR, Nagy E, Heise B (1981) Colloid & Polymer Sci 259: 789
16. Waring JR, Lovell R, Mitchell GR, Windle AH (1982) J Mat Sci 17: 1171
17. Krogh-Moe J (1956) Acta Cryst 9: 951
18. Narten AH (1977) J Chem Phys 67(5): 2102
19. Lovell R, Mitchell GR, Windle AH (1979) Acta Cryst A35: 598
20. Lorch E (1969) J Phys Chem 2: 229
21. Stewart CW, von Frankenberg CA (1967) J Polym Sci A-2 5: 623
22. Bunn CW, Howells ER (1954) Nature 174: 549
23. Quinn FA Jr, Robert DE, Work RN (1951) J Appl Phys 22: 1085
24. Marx P, Dole M (1955) J Am Chem Soc 77: 4771
25. Klug AK, Franklin RE (1958) Faraday Disc Chem Soc 25: 104
26. Hyndman D, Origlio GF (1960) J Appl Phys 31: 1849
27. Clark ES, Muus LT (1962) Zeitschr Kristall 117: 108
28. Clark ES, Muus LT (1962) Zeitschr Kristall 117: 119
29. Natarajan R, Davidson T (1972) J Polym Sci Polym Phys Ed 10: 2209
30. Weeks JJ, Clark ES, Eby RK (1981) Polymer 22: 1480
31. Farmer BL, Eby RK (1981) Polymer 22: 1487
32. Yamamoto T (1985) J Polym Sci Polym Phys Ed 23: 771
33. Albrecht T, Elben H, Jaeger R, Kimmig M, Steiner R, Strobl G, Stühn B (1991) J Chem Phys 95(4): 2807
34. Brown RG (1964) J Chem Phys 40: 2900
35. de Santis P, Giglio E, Liquori AM, Ripamonti A (1963) J Polym Sci A1: 1383

Received January 17, 1992

PVA-Iodine Complexes: Formation, Structure, and Properties

Keizo Miyasaka
Department of Organic and Polymeric Materials, Tokyo Institute of
Technology, Meguro-ku, Tokyo 152, Japan

The PVA-Iodine complexes formed in PVA films soaked in iodine-KI aqueous solutions without boric acid are studied from the structural point of view. First, iodine soaking at comparatively low iodine concentrations is studied where iodine sorption takes place mostly in the amorphous phase. There, our interest is concentrated on the following problems: What happens in PVA films during iodine soaking? How does the solid structure of PVA films affect the formation and properties of the complex? How does the chain extension affect the complex formation and properties? What is the structure of the complex formed in the amorphous phase? Then iodine soaking at high iodine concentrations is studied where iodine sorption takes place in the crystal phase as well as in the amorphous phase.

1 Introduction

Iodine makes blue colored complexes with many substances such as starch [1, 2] nylon-6 [3], poly(vinyl pyrrolidone) [4], poly(vinyl alcohol) PVA [5, 6]. From the application point of view, the blue PVA-Iodine complex is the most important among them, for it is widely used for film polarizers [7, 8]. The polarizers are prepared by soaking PVA films in a solution of iodine and potassium iodide (KI) with boric acid, and subsequent drawing to cause the high degree of uniaxial orientation [9]. The most important application of the polarizer is for liquid crystal display systems.

The blue PVA-Iodine complex was first found by Herrmann et al. [5] who first synthesized PVA, and by Staudinger et al. [6] as early as 1927. Since then, the complex has attracted the interest of many researchers, and many studies have been done from different points of view. Some studied the complex formed in PVA solutions, while others studied the complex formed in bulk (film and fibers) PVA. Some concentrated their attentions on the effects of the PVA molecular structure on the complex formation, i.e. the 1,2 glycol hetero structure content of the chain backbone [10–12], the stereoregularity [10, 11, 13–15], the degree of saponification [16–18] and the chemical modification such as formalization [18–20].

Murahashi and his coworkers [10, 11] showed that the 1,2 glycol hetero linkage has an inhibitive effect on the blue complex formation and estimated the sequence length of 1,3 glycol (regular) units, required for the complex formation of atactic PVA in a solution, to be about 120. Recently the use of PVA with a smaller hetero structure content than that of conventional PVAs was recommended for production of the polarizer [12]. As to the stereoregularity, Imai and his coworkers [13] showed that at room temperature, a syndiotactictly rich PVA, obtained by polymerization at low temperature, makes the blue complex in a solution whose PVA concentration is too low for the ordinary atactic PVA to form the complex. Murahashi and his coworkers [10, 11] also showed that the syndiotactictly rich PVA has a high ability of complex formation, and makes the wave length at the visible light absorption maximum, λ_{max}, shifts to larger values. On the other hand, a PVA with a mole fraction of 5% of 1,2 glycol content and 70% isotacticity makes no complex in the solution even at a temperature low enough for normal PVA to form the complex [10].

Matsuzawa and his coworkers [14] also showed the high degree of complex formation possible with syndiotactictly rich PVA, in a series of studies. The same was shown by our group [15]. It should be noted that the high capacity for complex formation of the syndiotactictly rich PVA implies that the PVA chain segments participating with the complex may be those of the syndiotactic configuration. Hayashi and his coworkers [16–18] studied the red complex formation of partially saponified poly(vinyl acetate) PVAc. According to them, an approximately 88% saponified PVA, which is as soluble in water as PVA, forms the same red iodine complex as PVAc before saponification [16]. This indicates that the saponification takes place irregularly with the remaining PVAc sequences long enough to form the same red PVAc-Iodine complex as pure PVAc does. They also studied the effect of formalization of PVA to show that the partial formalization to a mole fraction of about 20% enhances the formation of the blue complex [18–20]. It is interesting that the effect of formalization is also apparent in the complex formation of partially saponified specimens: a specimen which only forms the red complex becomes capable of forming the blue complex as well as the red one after partial formalization [18]. Our group [21] showed that formalization of PVA increases the stability of the complex against desorption of iodine, although it restricts the complex formation remarkably, as will be shown later.

The PVA-Iodine complex formation has also been widely studied in relation to the preparation conditions such as the concentrations of iodine and PVA in the case of complex formation in solution, the iodine concentration in the case of complex formation in bulk PVA by soaking PVA films, and the temperature of complex formation. As to the concentrations, the following results are always obtained: the higher the concentrations of PVA and iodine in solution, the greater the amount of complex formed in the solution [22–24]. The increase of the iodine concentration in soaking solutions also increases the amount of the complex formed in PVA films [25]. On the other hand, a decrease in temperature is favorable for complex formation [6, 10, 26]. The effect of the addition of boric acid on complex formation is also important, for it remarkably enhances complex formation [23, 27–29] and therefore is widely used in the production of the polarizers. The effect of boric acid on complex formation is so great that the effect of stereoregularity of the PVA chains almost disappears in the presence of large amount of boric acid [23]. The effect of boric acid is also apparent in the complex formation of partially saponified PVA in solutions: Addition of boric acid results in a large amount of the blue complex in addition to the red one in a solution in which only the red complex forms without the addition of boric acid. Boric acid is believed to make a bridge between PVA chains [30] and this must decrease the freedom of chains and give a higher stability to the complex. Recently a new model for the PVA inter-chain bridge by boric acid has been proposed by Nomura and his coworkers [31]. The effect of addition of orthotelluric acid on the PVA-Iodine solution was also studied [32].

The structure and properties of the complex have also attracted the interest of many researchers. West [27, 33] found that soaking uniaxially drawn PVA films with high molecular orientation in iodine results in strong uniaxial positive dichroism and that the films have the characteristic visible ray absorption spectrum and X-ray diffraction with a layer-like meridional peak corresponding to a spacing of 310 pm peak. He reached the conclusion that linear polyiodines in which the atoms form a linear lattice with a repeat distance of 310 pm oriented parallel to the direction of draw of the specimen, i.e. the direction of molecular orientation. It should be noted that the X-ray layer pattern extending parallel to the equator indicates that the iodines form no three dimensional crystal but a one dimensional lattice. As the repeat distance of the PVA chain is 250 pm, the 310 pm peak is doubtlessly due to the iodine lattice. Haisa and his coworkers [34] analyzed the breadth of the layer-like meridional peak to find that the polyiodine reflecting the peak consists of more than 15 iodine atoms. It should be noted that this length of the lattice is long enough to be compared with the thickness of an amorphous layer[1] which exists sandwiched between stacked crystallites.

Resonance Raman spectroscopy, recently developed, has made a great contribution to the investigation of the polyiodines participating in the complex. The identification of the polyiodine by resonance Raman spectroscopy is made using a standard material whose polyiodine structure is well established. Heyde et al. [35] and Inagaki et al. [36] found that the resonance Raman spectrum of the blue PVA-Iodine complex forming in the solution is similar to that of the Amylose-Iodine complex[2], and assumed that I_3- mode polyiodines form in the complex. At that time, it was believed that I_3- prevails over others in the Amylose complex. However, Teitelbaum et al. [37] showed that it was the I_5- mode complex which prevails over other modes in the Amylose-Iodine complex which is as popular as PVA-Iodine complex.

Our resonance Raman spectrum study on iodinated blue films showed that both I_5- and I_3- mode polyiodines form in the films, and the former prevails over the latter in the amorphous phase of PVA. On the other hand, the spectra of the specimens soaked at high iodine concentrations show that the I_3- mode complex forms in the crystalline phase. Then the color of the films is not blue but purple or almost red.

Yokota and his coworkers [26, 42–44] made systematic studies on the temperature dependence of the stability of the blue PVA-Iodine complex, using visible ray absorption spectrum. Attention was paid to the structure of the polyiodine and its mechanisms of formation and decomposition. With regard to the composition of the polyiodine, they carried out a very interesting experiment [42]: First the polyiodines participating in the complex were separated from others in an aqueous PVA-Iodine-boric acid solution in which the blue complex formed, using anion exchange resin. Then the molar ratio I_2/I^- of the separated polyiodine was estimated, with the result that the ratio is approximately 2. They concluded that the polyiodine is I_5- with linear configuration and or I_3- with distorted configuration.

The visible light absorption spectrum of iodinated PVA was also studied by many authors [19, 23, 26, 41] to find that it has an absorption peak whose maximum wave length λ_{max} is in a range from about 520 nm to 700 nm, depending on the preparation conditions such as iodine concentration and addition of boric acid [23]. The λ_{max} also varies depending on the molecular structure of PVA specimens, i.e. stereoregularity [10, 11], and acetalization [19, 45]. The degree of hydration and extension of swollen PVA films also change the λ_{max}, as will be shown later [25, 46, 47]. The change in the λ_{max} is

---

[1] For example, in a PVA film whose crystallinity and long spacing are 0.4 and 10 nm respectively, the thickness of the amorphous layer is approximately 6 nm. Although this is only an example, it may be useful to take these dimensions of the structure into consideration.

[2] Many studies have also been made on the structure and properties of the Amylose-Iodine complex [37–40]. Zwick [41] published a concise review of some important features of the Amylose-Iodine complex.

often accompanied by the change in the amount of complex formed. For example, the change in the equilibrium degree of swelling of PVA films causes changes both in the λ_{max} and the amount of complex formed, as will be shown later. As the λ_{max} is characteristic of the mode of the polyiodine in the complex, a remarkable shift of λ_{max} is an indication of some changes in the modes of complex formed or in their fractions of the amounts.

The conformation of the PVA chain segment in the complex is also one of the important problems of the complex structure. Zwick [23] studied the effect of boric acid on the complex formation using a spectrophotometric titration technique, and proposed a structural model of the PVA-Iodine complex in which a single chain goes round a polyiodine helically, which is basically the same structure as that of the Starch-Iodine complex [38]. He supported the idea that the helix mode complex tends to be associated, on the basis of the irreversibility of the complex formation, with the PVA concentration. Inagaki et al. [36] supported Zwick's model on the basis of similarity of Raman spectra between PVA-Iodine and Amylose-Iodine complexes. It should be noted that Raman spectra give important information on the mode of polyiodines but only poor information about the mode of PVA chain conformation. On the other hand, Rundle and his coworkers [38] proposed another model in which a polyiodine is supported by several PVA chains, based on the fact that in the experimental result the complex formation requires a comparatively high PVA concentration. Tebelev and his coworkers [24] found a difference in the concentration dependence of visible light absorption between starch and PVA complexes: in the case of PVA, the absorption changes in a concave mode with increasing PVA concentration, but in a convex mode in the case of starch. They supposed that the difference in the concentration dependence of absorption is due to the difference in the modes of chain conformation between the two complexes, supporting Rundle's model. Due to these circumstances, a decisive conclusion has not yet been drawn on the PVA chain conformation in the PVA-Iodine complex.

We have to pay attention to the iodine concentration at which the specimens are soaked when we discuss the structure and properties of complexes in iodine soaked PVA films. This is due to the fact that iodine only penetrates into the amorphous phase at low iodine concentration, while it penetrates into the crystalline phase as well as the amorphous phase at higher concentrations. Hess and his coworkers [48] found that a new X-ray diffraction peak appears when PVA fibers absorb more than 12% iodine, which is caused by the penetration of iodine into the crystal phase. As expected, iodine sorption in the different phases of PVA must cause different modes of complex.

In this manuscript, the author would like to review the structural studies, made by his group, on the PVA-Iodine complexes formed in PVA films soaked in iodine-KI solutions. This review starts with what happens during soaking of PVA films in iodine, discusses next the effects of structure and extension of PVA films on the formation and properties of complexes, and ends with the structures of the complexes.

2 General Features of Iodine Soaking of PVA Films

2.1 Structure of Water-Swollen PVA Films

It is interesting to consider the structure of PVA films in the water-swollen state
and then proceed to discuss complex formation by iodine soaking.

Our group [46, 49] proposed a structural model for solution-cast PVA films
to explain the effects of annealing on the crystallinity and the degree of water
swelling, as shown in Fig. 1. According to our results, after annealing at high
temperatures, a film cast from aqueous solution remarkably decreased the
equilibrium degree of water swelling at room temperature, while the crystallinity
hardly increased at all, as detected by X-ray diffraction, IR spectrum, density
and DSC measurements. This small effect of annealing on the crystallinity may
be partly due to the fact that the specimen was an as-cast undrawn film, for a
detectable increase in the crystallinity is usually expected for drawn PVA
specimens. Our model consists of a microfibrillar network, and each microfibril
consists of crystallites and amorphous layers successively stacked in series

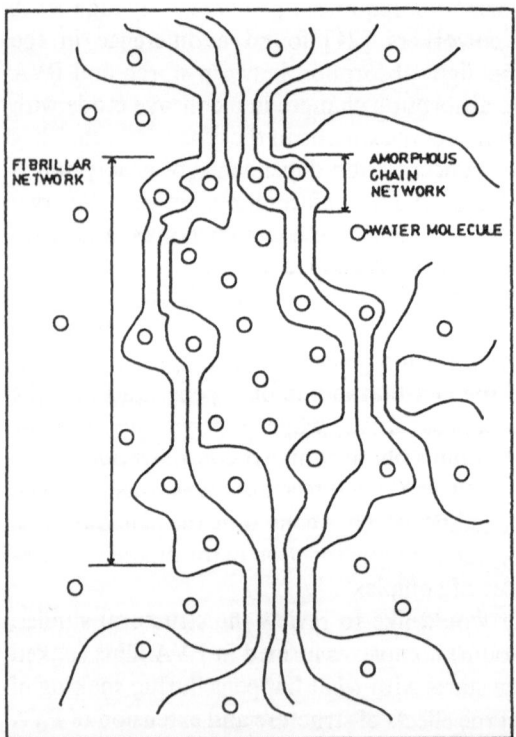

Fig. 1. A schematic representation of the double-network structure

forming a so-called long spacing, as detected by small angle X-ray scattering (SAXS). In the amorphous layer, a chain network is formed by tie chains whose both ends are fixed at the interfaces of neighboring crystals. As there are two networks in our model, i.e. chain network and fibrillar network, it is called a double network model. Swelling of a film is caused by water sorption in both the amorphous layers and the interfibrillar spaces. The remarkable decrease in the degree of swelling by annealing must be due to the developments of both networks [3], i.e. the adhension of fibrils, and the coagulation of tie chains in a so small scale that no noticeable increase in the crystallinity is detected.

2.2 Change in the Degree of Swelling During Iodine Soaking

A PVA-Iodine complex forms, when PVA films are soaked in iodine-KI aqueous solutions with iodine concentrations higher than the threshold value. However, some changes other than the iodine sorption followed by complex formation, occur in the films, during soaking such as those shown in Fig. 2, i.e. [50] the decreases in the volume (degree of swelling) and the long spacing, and the increase in Young's modulus. In this case in Fig. 2, no iodine sorption takes place in the crystalline phase, and the complex forms only in the amorphous phase, as the iodine concentration is not high. The changes in the SAXS intensity profile with iodine soaking is shown in Fig. 3.

The weight of swollen PVA film rapidly increases due to iodine sorption during the first few minutes, and then decreases following the release of water. The volume contraction of a swollen specimen and the shrinkage of the long spacing take place comparatively rapidly and end in an hour. The formation of the complex takes place after the iodine sorption, as will be shown by the time effect on the visible ray absorption spectrum in Fig. 4. It is natural that the rate of the complex formation is much slower than that of the iodine sorption because formation of polyiodines and coagulation of PVA segments are necessary for the complex formation. It is interesting to note that the relative decrease in the long spacing with iodine soaking is larger than that of volume of specimens. If an affine mode shrinkage occurs, the shrinkage of the long spacing should be equal to the cube root of the contraction of the volume, i.e. $(L/L_0) = (V/V_0)^{1/3}$, where L and V are the long spacing and volume, respectively, and suffix 0 shows values at the equilibrium water-swollen state before iodine

---

[3] In the original paper [46], we supposed that since the swelling of the amorphous layers is necessarily affected by the crystallinity, annealing, accompanied by no noticeable increase in the crystallinity, must result in only a small change in the degree of swelling in the amorphous layer, and that it causes interfibrillar adhesion without an increase in the crystallinity, resulting in the development of fibrillar network. This necessarily decreases the water sorption in the interfibrillar space. Then the remarkable decrease in the degree of swelling of the specimen, induced by annealing, is possible without any noticeable increase in the crystallinity. We have now changed the consideration to that mentioned above in the text. However, we still believe in the usefulness of the double network structure.

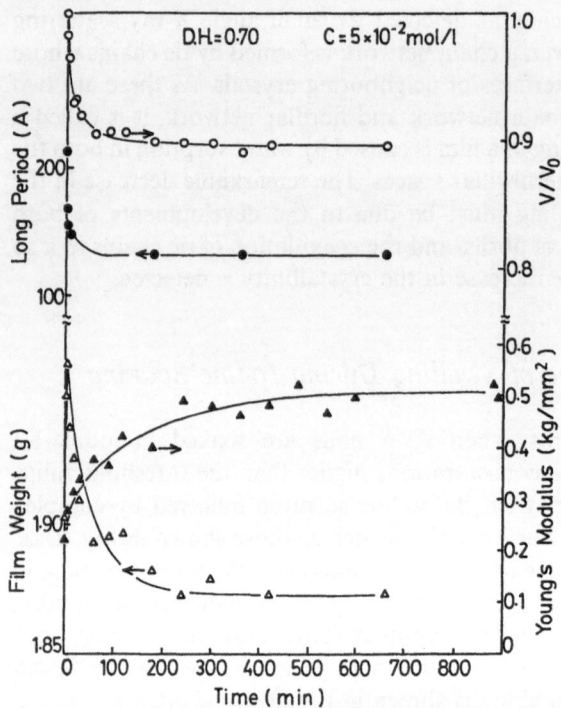

Fig. 2. The time dependency of volume, Young's modulus, long period and weight of the swollen PVA film during soaking in the iodine solution

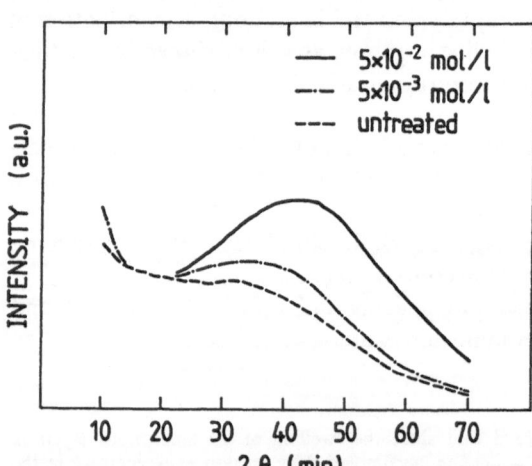

Fig. 3. Change in small angle X-ray scattering profile of the swollen PVA film of complexation

soaking. However, the results show that no affine mode shrinkage occurs. This may be well related to the double network structure just cited above. The shrinkage of the long spacing is caused by the contraction of the amorphous layer due to the shrinkage of the chain network, while the volume contraction is

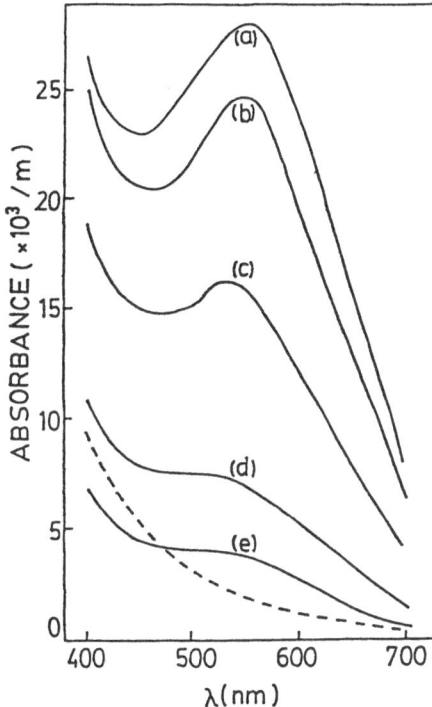

Fig. 4. Absorption spectra of iodine sorbed PVA films at various iodine concentrations: (a) 3.0×10^{-3} mol/l; (b) 2.5×10^{-3} mol/l; (c) 2.0×10^{-3} mol/l; (d) 1.5×10^{-3} mol/l; (e) 1.0×10^{-3} mol/l at 5 °C, together with the base line (2.0×10^{-3} mol/l, 10 min.) given by *dashed line*

related to the decrease in the degrees of swelling both in interfibrillar space and in the amorphous layer. It should be noted that if the two networks in our model have not undergone any change by iodine soaking, then an affine mode contraction must be necessary. Thus the results show that the interfibrillar network tends to deteriorate to some extent, and in other words, some loosening of the network is induced by iodine sorption, making the swelling of the interfibrillar network easier than before. This loosening may be on the same line as the result that iodine sorption makes air dried PVA more ductile, increasing the maximum draw ratio remarkably [51].

It must be that iodine is sorbed to be much condensed in the amorphous layers, and a substantial number of which participate in the complex formation, while in the interfibrillar space the soaking solution may take up part of the water which existed before soaking, and complex formation is hardly expected there. The increase in the SAXS intensity of the long spacing peak in Fig. 3 seems to support the condensation of the iodine in the amorphous layers. This means that most of the complex form in the amorphous layers with participation of tie chains. The change of Young's modulus of the specimens takes place so slowly that it lasts for 10 hours, which is related to some reinforcement of the chain networks due to the release of water and complex formation.

3 Effect of the PVA Film Structure on Complex Formation

It is an interesting problem how the solid structure of PVA films affects complex formation. It is well known that the solid structure of crystalline flexible polymers is not homogeneous but very complicated.

The heterogeneity is mostly related to the coexistence of the crystal and amorphous phases, which makes the characterization of the structure difficult, although the estimation of the fraction of each phase is comparatively easily made: the fraction of the crystal phase is called the crystallinity, the estimation methods of which have been already established. It should be noted that the characterization of the structure concerned with the crystal phase is much more easily and reasonably made than that concerned with the amorphous phase. For example, the orientation of the crystal phase can be measured exactly by X-ray, while for the amorphous phase only the averaged value which is sometimes called Hermann's orientation function can be estimated with less accuracy than that for the crystal phase. As to the structure of the amorphous phase, the small angle X-ray scattering (SAXS) indicates that the amorphous phase forms sandwiched between successively stacked crystallites, making a so called long spacing. Because of the finding of polymer single crystals with the folded chain morphology [54], it is indisputable that the fringed micellar [53] structure is unrealistic for the morphology of crystalline polymers. However, it is still disputable what the amorphous chains are: many kinds of amorphous chains are figured [55] such as a loop chain whose both ends are fixed on the same crystal interface, and a cilium of the end of parent chain whose one end is fixed at the crystal interface, and a tie chain whose both ends are fixed on the interfaces of two crystals successively stacked. The last is substantially the same as that of the fringed micellar model, and is considered to play the most important role, among others, in the complex formation on which our interests are concentrated. The second kind of chain is not important because of the small number of these chains. At this moment, however, it is impossible to draw a quantitatively detailed figure of the amorphous phase which gives such informations as the fraction of each mode of chain, their molecular weight (length), and their degree of extension relating to their conformation. All of these are considered to be important structural factors affecting the formation of the complex.

Under these circumstances, the equilibrium degree of water swelling is useful as the factor characterizing the structure of the PVA films. This is because the degree of swelling is highly sensitive to the difference in the fine structure, as seen in the annealing of PVA films cast from solutions where a remarkable decrease in the degree of water swelling is caused, while the crystallinity remains almost unchanged [49]. Another reason for the use of the degree of swelling as the factor characterizing the structure of PVA films is that the complex formation with iodine is actually much affected by the degree of swelling, as will be shown shortly. The following remarks, however, may be noted: The complex seems to be formed mostly in the amorphous layers, as will be shown, although it is also formed in the crystalline phase when the iodine concentration of the soaking solution is high. However, the information about the swelling of the amorphous layer itself is not available because of the complexity of the swelling of PVA films, although the degree of swelling of the whole sample is easily measured. This is because swelling of PVA films seems to be caused not only by the sorption in the amorphous layer but also by sorption in the interfibrillar space, as suggested by the double network concept [46, 49]. In spite of these problems, the equilibrium degree of water swelling is still a useful and important factor characterizing PVA films in the study of complex formation.

In solution, the increase in the PVA concentration enhances complex formation [22–24]. If this is still the case in swollen films, the smaller degree of swelling is favorable for complex formation. Furthermore, the effect of extension of tie chains caused by swelling on the formation of complex is an interesting but difficult problem to solve. On the other hand, amorphous PVA chains must be under a constraint resulting mainly from the crystal interfaces, which may have some inhibitive effect on complex formation. The constraint must be more severe in specimens with a lower degree of swelling. These considerations make us expect the complicated features of the effects of the degree of swelling on the complex formation. The degree of hydration (D.H.) of PVA films, defined as the volume fraction of water in the swollen state and calculated from the equilibrium degree of swelling, is often used to characterize the swelling properties, as it is in this case.

Figure 4 shows visible light absorption spectra of specimens soaked in iodine-KI aqueous solution with different iodine concentrations for 24 hours, which was long enough for equilibrium to be reached. A broad absorption peak with a maximum at about 600 nm, assigned to the PVA-Iodine complex, intensifies with increasing iodine concentration. The effect of iodine concentration on the D.H. dependence of λ_{max} is small in this case because of the narrow span of the iodine concentration. The λ_{max} depends on the structure of the complex, particularly on the number of atoms in polyiodine participating in the complex, while the absorbance at the λ_{max} is approximately proportional to the amount of complex formed. Figures 5 and 6 show the effect of the degree of hydration of PVA films on the λ_{max} as function of iodine concentration and temperature, respectively. These figures indicate that the λ_{max} values markedly shift to the longer side with increasing D.H., and that the temperature effect on λ_{max} is also remarkable, although the effect decreases with increasing temperature.

As will be discussed in more detail later, in this case the I_5- and I_3- mode complexes form, although the former prevails over the latter, and the λ_{max} of the former is larger than that of latter. Thus the shift of λ_{max} to the long side, caused by increased D.H. and decreased temperature, is attributed mainly to the gradual increase in the fraction of the I_5- mode complex.

Figure 7 shows the absorbance Q_c at λ_{max} of specimens soaked at different iodine concentrations C of the soaking solutions. In this case, the Q_c is roughly proportional to the equilibrium amount of the complex formed. According to

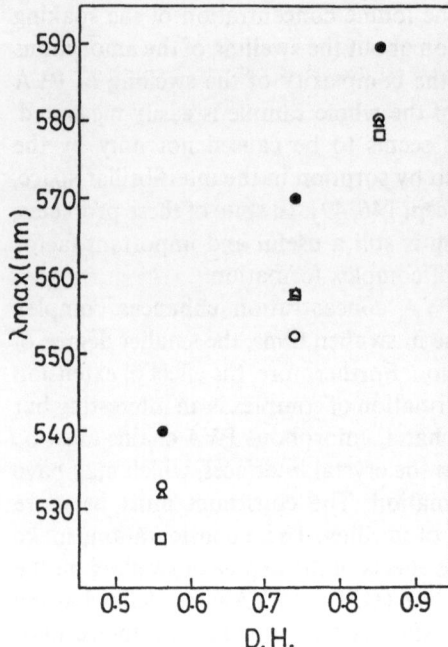

Fig. 5. The absorption maximum λ_{max} as a function of D.H. at various iodine concentration: ($\bullet$) 4.5×10^{-3} mol/l; ($\bigcirc$) 4.0×10^{-3} mol/l; ($\triangle$) 3.5×10^{-3} mol/l; ($\square$) 3×10^{-3} mol/l at 30 °C

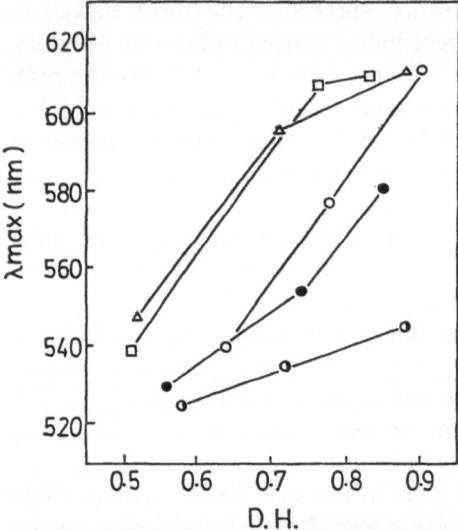

Fig. 6. The absorption maximum λ_{max} as a function of D.H. at various temperatures: ($\square$) 5 °C; ($\triangle$) 15 °C; ($\bigcirc$) 25 °C; ($\bullet$) 30 °C; ($\circleddash$) 45 °C

Fig. 7, the amount of complex increases with increasing iodine concentration at both 5 °C and 45 °C. It was found that Eq. (1) can describe the relation between C and Q_c in Fig. 7.

$$Q_c = (KC)^A \tag{1}$$

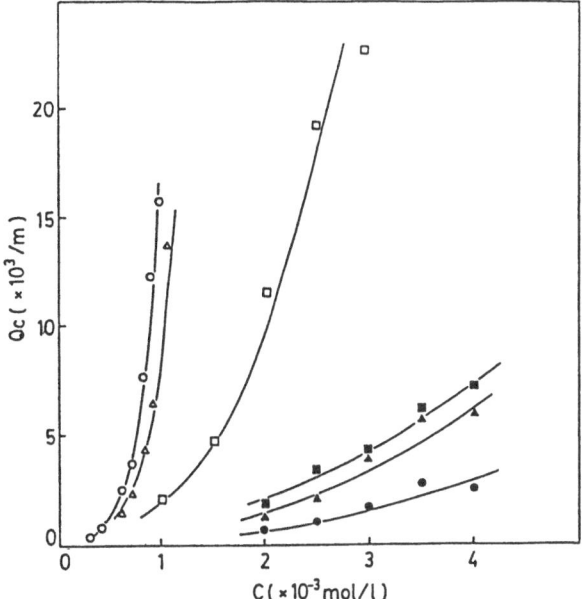

Fig. 7. The relation between the amount of complex Q_c and the iodine concentration C_I in the solution as a function of D.H.: (○) 0.83; (△) 0.76; (□) 0.51 at 5 °C, (●) 0.88; (▲) 0.72; (■) 0.58 at 45 °C

where C is the iodine concentration of soaking solution in mol/l and K is a constant equal to 1.2×10^4. This means that A is the determining factor of Q_c at a given soaking condition. Figure 8 shows the parameter A as a function of D.H. at different temperatures. As the parameter A is for the equilibrium state, the temperature dependence of A is free from the temperature effect on the rate of complex formation. According to Fig. 8, the parameter A increases almost linearly with increasing D.H. at all temperatures, although the slope decreases with increasing temperature and becomes even slightly negative at 45 °C. This indicates that less thermal agitation of chains is favorable for complex formation.

In homogeneous networks such as that of crosslinked rubber, the equilibrium degree of swelling q_m depends on the molecular weight of the network chain M_c, satisfying the following relation for a given rubber network-solvent system [56],

$$M_c \propto q_m^{5/3} \qquad (2)$$

This relation is not applicable to our heterogeneous network. We may, however, assume that the average molecular weight of tie chains may be larger in larger D.H. films than in smaller D.H. films as discussed in Sect. 2.1. This makes us suppose that the parameter A may depend on the freedom of tie chains in the

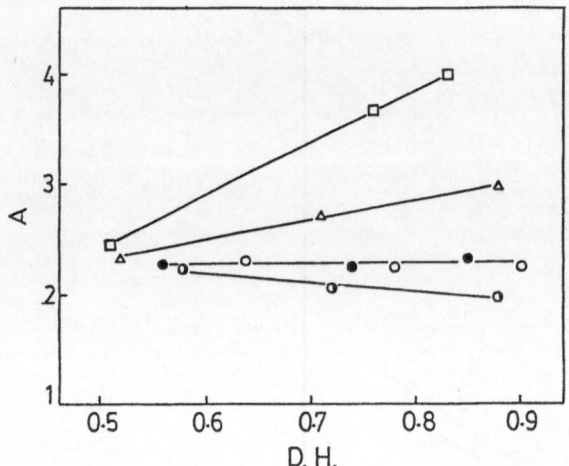

Fig. 8. Parameter A as a function of D.H. at various temperatures: (□) 5 °C; (△) 15 °C, (○) 25 °C; (●) 30 °C; (○) 45 °C

swollen state relating to their molecular weight. The longer chains are favorable for complex formation, unless thermal agitation is too intense.

The decrease of the parameter A with increasing D.H. observed at 45 °C is interesting when compared with the following results on the thermal stability of the complex in the soaking solution: after being soaked in a solution with a given iodine concentration at 5 °C until the equilibrium amount of complex is formed, the solution with the specimen was heated stepwise and kept at each temperature for 30 min, and the amount of complex was measured. The results are shown in Fig. 9. According to Fig. 9, the amount of complex begins to decrease at 20 °C, and the decrease continues until the complex disappears completely at 75 °C and 60 °C in the specimens with D.H. = 0.51 and 0.83, respectively. This means that the thermal stability of the complex is higher in the specimen with smaller D.H. than in the one with larger D.H. The λ_{max} of the specimens remains almost unchanged during heating, as shown in Fig. 9, which is in contrast to the results shown in Fig. 4 that λ_{max} decreases with increasing soaking temperature at a given D.H.

The decrease in the amount of complex with increasing temperature in Fig. 9 is qualitatively in accordance with the temperature effect on the complex formation shown in Figs. 7 and 8, except that the temperature effect appears even below 20 °C in Fig. 5 while in Fig. 9 the decrease in the amount of complex only begins at 20 °C. This exception may arise from the dependence of the stability of the complex on the soaking temperature. The stress relaxation data for these specimens measured in water, shown in Fig. 10, are useful to study the reason why the amount of complex begins to decrease at about 20 °C. The data in Fig. 10 were obtained under 20% extension and the same heating conditions as in Fig. 9. Although the stress values are quite different between two specimens

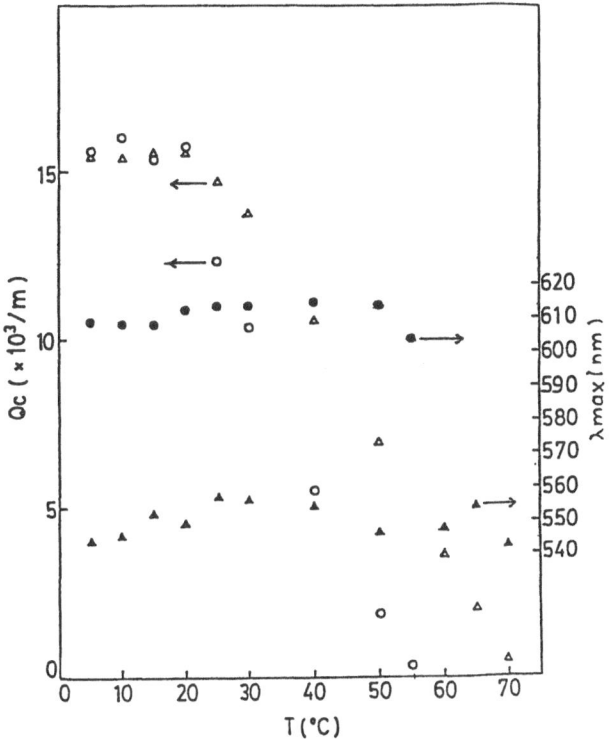

Fig. 9. Decomposition behavior of iodine-PVA complex for different D.H.s: (○, ●) 0.83, 1×10^{-3} mol/l; (△, ▲) 0.51, 2.5×10^{-3} mol/l as a function of temperature

with different D.H.s, the shapes of their stress relaxation curves are very similar to each other. The comparison of Fig. 9 with Fig. 10 indicates that the thermal stability of the complex is closely related to the thermal mobility of the PVA chains in water. Figure 10 indicates that water swollen PVA has a transition point at 20 °C which affects the chain mobility so much that it may be assigned to T_g.

In the case of Poly(vinyl acetate) -Iodine complex, it has been reported that the complex suddenly disappears when the temperature is above the T_g of the polymer [57, 58]. In the PVA-Iodine complex, the amount of complex begins to decrease at the 20 °C transition, followed by a gradual decrease in a wide span of temperature. It should be remarked that this 20 °C transition, demonstrated by the beginning of the decrease in the amount of the complex in PVA films may be the same as that observed by Yokota et al. [26] in aqueous PVA/iodine-KI boric acid solutions. They studied the thermal stability of the blue complex in the solution to find that dissociation of polyiodine begins at 10–15 °C. Although our case is that of PVA bulk without boric acid, while their case is that of solution with boric acid, the difference is not substantial because the complex is

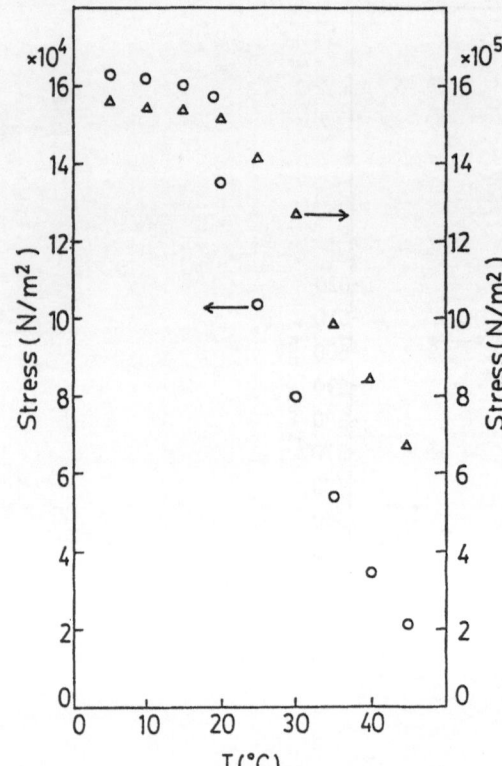

Fig. 10. The stress relaxation at a constant length of PVA films for different D.H.s: (○) 0.83; (△) 0.51 as a function of temperature

formed by PVA chains coexisting with a large amount of water. The important conclusion drawn from this result is that dissociation of the complex is induced by the thermal movement of PVA chains, both in bulk and solution. This again demonstrates that the complex formation is controlled by both the freedom and the thermal movement of PVA chains. Figure 9 also shows that the melting point of the most stable complex formed in the specimen of D.H. = 0.51 is about 75 °C and that of the D.H. = 0.83 specimen is 60 °C, while that of the least stable complex in these specimens is about 20 °C. This wide span of the melting point of the complex must be closely related to the heterogeneous states of swollen PVA chains.

4 Effects of Extension of PVA Films on Complex Formation

Extension of PVA films remarkably enhances the PVA-Iodine complex formation [46]. When brown colored PVA films (in which iodine is sorbed but no complex is formed because the iodine concentrations are too low), are stretched in the soaking solution, they change the color from brown to blue during extension, showing the complex formation. The formation of the PVA-Iodine complex induced by extension must be due to the strain energy stored in PVA chains.

We [47] further studied quantitatively the strain-induced complex formation in PVA films in dilute iodine solutions whose iodine concentration is lower than the threshold required for the complex formation. We were interested in the effects of degree of hydration D.H. of PVA films and the iodine concentration of the soaking solutions on the strain-induced complex formation. PVA films were stretched in iodine-KI soaking solutions whose iodine concentration was in the range of 2×10^{-4}–9×10^{-4} mol/l. No boric acid was added to the solution. No complex forms in these solutions, and therefore films remain brown in color before extension. However, when stretched in the solution the color turns to blue at the points indicated by arrows on the stress-strain curves shown in Fig. 11, which shows the beginning of the formation of the complex. The strain-induced complex formation is also shown by the visible ray absorption spectra measured before and after extension in a solution of 3×10^{-4} mol/l iodine concentration at 30 °C as shown in Fig. 12. A broad peak of the complex with λ_{max} at about 700 nm appears at 300% extension in the solution, while no peak appears before extension.

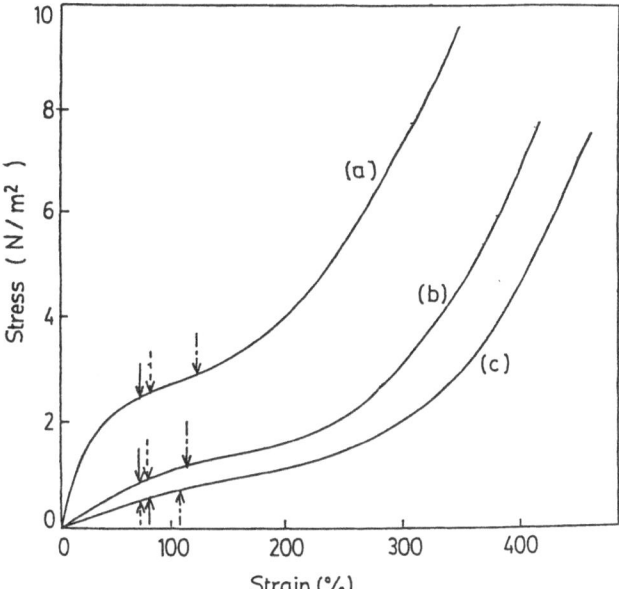

Fig. 11. Stress-strain curve of iodine-sorbed PVA films of D.H.; (a) 0.64, (b) 0.77, (c) 0.85 at various iodine concentrations: (– – –), 3×10^{-4} mol/l; (- - - -), 6×10^{-4} mol/l; (———), 9×10^{-4} mol/l at 30 °C

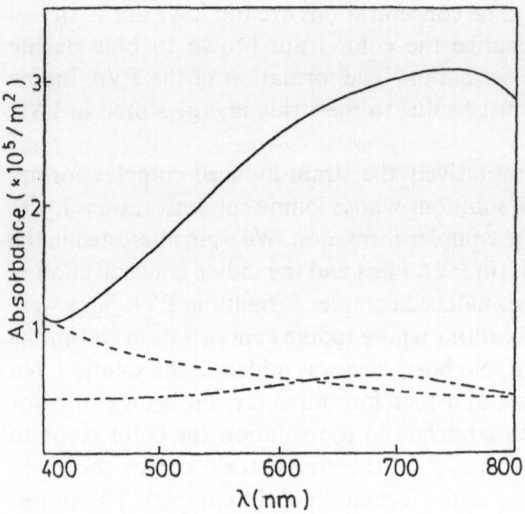

Fig. 12. Absorption spectra of iodine-sorbed PVA films on extension in 3×10^{-4} mol/l of iodine at 30 °C: (——), 300% extension in the solution; (- - - -), 0% extension in the solution; (- - -), soaked in the solution after 300% extension

Figure 13 shows the absorbance at λ_{max} of a specimen of D.H. = 0.84, soaked in 3×10^{-4} mol/l iodine concentration, as a function of the extension at 30 °C. This absorbance is proportional to the amount of complex formed. The amount of the complex increases successively with extension after passing the point indicated by an arrow on the stress–strain curve. All the swollen PVA films used are so highly elastic that the strain is released when the extension is not more than about 100%. The complex disappears after removal of the extension force when strain is released. This reversible strain-induced complex formation is explained by the increased free energy of PVA chains due to the extension.

Let g_c and g_s be the free energy of the complex and non-complex phases, respectively. The g_s depends not only on the concentrations of iodine and PVA, and temperature but also on the conformational energy of the PVA chains. Let T_d be the transition temperature (melting point) where g_c is equal to g_s, and the non-complex phase is stable above T_d. At T_d, the following equation must be satisfied.

$$T_d = \Delta H_{comp} / \Delta S_{comp} \tag{3}$$

where ΔH_{comp} and ΔS_{comp} are the differences in enthalpy and entropy between complex and non-complex phases. Thus T_d depends on the iodine concentration and the state of PVA chains which is related to the fine structure of PVA films as characterized by D.H. When T_d is lower than the soaking temperature, no complex forms, as is the case with very low iodine concentrations. When PVA

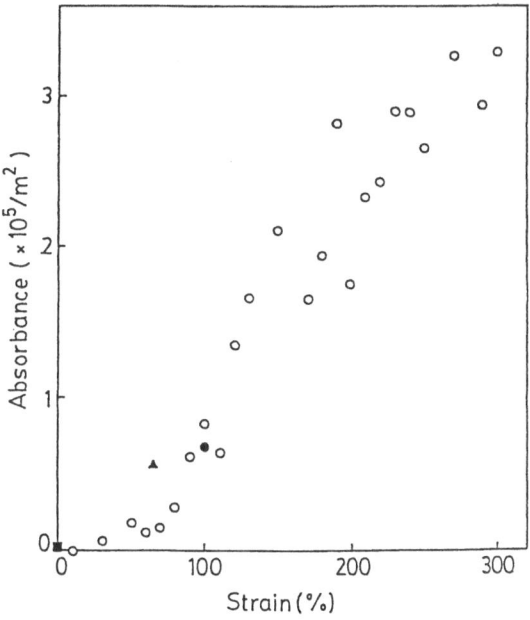

Fig. 13. Absorbance as a function of strain at 30 °C, together with the absorbance on recovery: (●), recovery from 300% extension; (▲), recovery from 200% extension; (■), recovery from 100% extension

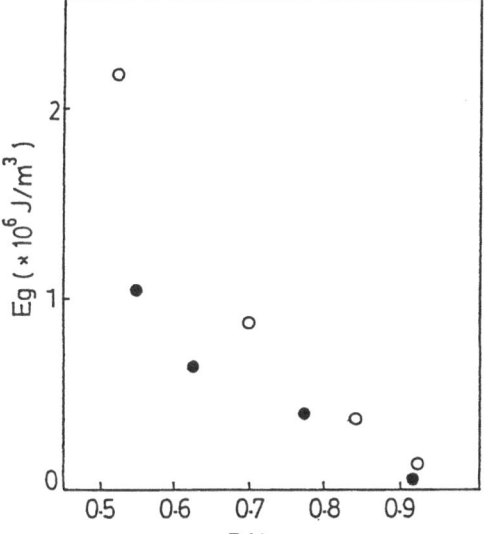

Fig. 14. Total energy per unit volume of gel as a function of D.H., at 30 °C (○) and 5 °C (●)

chains in the amorphous phase are extended to increase g_s, mainly due to the decrease in their conformational entropy, T_d necessarily shifts to a high temperature. When T_d is above the soaking temperature, the complex forms. The explanation of the enhancing effects of boric acids [23, 27–29]

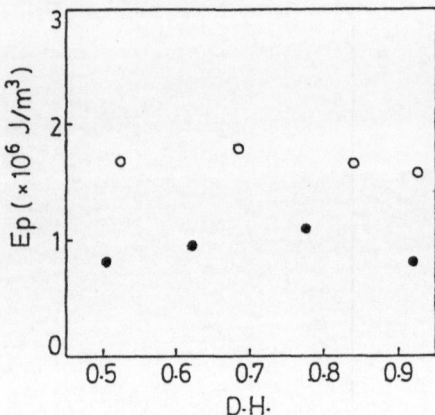

Fig. 15. Strain energy on recovery per unit volume of polymer as a function of D.H., at 30 °C (○) and at 5 °C (●)

and formalization [18–20] must also be the decrease of entropy of the PVA chains.

The strain energy required for complex formation, estimated from the stress–strain curves (Fig. 14) with the arrows pointing at the beginning of the complex formation, depends greatly on the D.H. of the PVA films as can be seen in the figure. However, when the energy is standardized by unit volume of polymer, the effect of the D.H. becomes small, as seen in Fig. 15, although the effects of temperature and iodine concentration are still remarkable. It should be noted that these results of strain-induced complex formation strongly suggest that the extended conformation of PVA chains is favorable to complex formation, implying that the conformation of the chain in the complex may be that of the extended type.

5 Structure of the PVA-Iodine Complex Formed in the Amorphous Phase of PVA Films

The X-ray diffraction diagrams of iodinated drawn PVA films show a meridional-layer-like peak of about 310 pm spacing [27, 33], indicating the existence of linear polyiodines oriented parallel to the draw direction of the films. This sort of polyiodine must be formed even in undrawn PVA specimens, although no X-ray intensity concentration on the meridian is observed due to a lack of preferential orientation. In this case, the important problems concerning the structure of the complex which remain unsolved are (1) the number of iodine atoms forming a linear polyiodine participating in a complex and (2) the conformation of the PVA chains participating in a complex. As to the first problem, resonance Raman spectrum study [35–37] and the stoichiometric analysis of iodines participating in the complex are very useful [42–44]. As to the second problem, Zwick [23] proposed a model of a helical conformation of the PVA chain surrounding a polyiodine, as in the amylose-iodine complex. Zwick's model was supported by Inagaki et al. [36] with Raman data on PVA-Iodine and Amylose-Iodine complexes which were very similar to each other. On the other hand, Rundle et al. [38] and Tebelev et al. [24] suggested a model of several PVA chains with an extended conformation surrounding a polyiodine, although they showed no concrete idea of the number of chains and the molecular packing state in the complex.

5.1 Structure of Polyiodine from the Resonance Raman Spectrum

Figure 16 shows the Raman spectra, obtained with a 514.5 nm excitation of an Argon ion laser, for the specimens soaked at different iodine concentrations. Two Raman shifts appear at 108 and 164 cm^{-1} for both specimens, although the intensity of the latter is much stronger than that of the former. There are no substantial differences among the spectra of specimens treated at different iodine concentration. The 164 cm^{-1} and 108 cm^{-1} peaks are assigned to the I_5- and I_3- mode polyiodines, respectively[4]. According to Teitelbaum et al. [37], symmetric I_3- species have a strong single Raman band due to the symmetric stretching vibration in a region from 108 to 118 cm^{-1}, depending on the structure of the I_3- species.

On the other hand, according to Ref. [59] the linear I_5- species has a strong band at around 160 cm^{-1}: the frequency varies depending on the I_5- symmetry. The resonance Raman spectra in Fig. 16 indicate that both I_3- and I_5- polyiodines are formed in the specimens soaked at low iodine concentrations. Although, Raman data give no exact information about the fraction of I_5- and I_3- mode polyiodines, the stronger intensity of the I_5- mode peak indicates that the I_5- mode polyiodine prevails over the I_3- mode one. This is supported by visible light absorption and stoichiometric analysis as well as by the Raman data already cited above.

5.2 A Model of the PVA-Iodine Complex

We proposed a model of the I_5- complex as shown in Fig. 17. Our model of the complex formed in the PVA amorphous phase has the structure of four syndiotactic PVA segments with the extended conformation surrounding a

---

[4]In Ref. [50], we concluded from Fig.16 that both the 108 cm^{-1} and the 164 cm^{-1} peak are assigned to the I_5- mode polyiodine. However, the conclusion has now been revised, because the revised identification is more reasonable on the bases of these results as just cited. The resonance Raman spectra of specimens treated at high iodine concentrations also support that the 108 cm^{-1} peak should be assigned to the I_3- mode polyiodine, as will be shown in the next section.

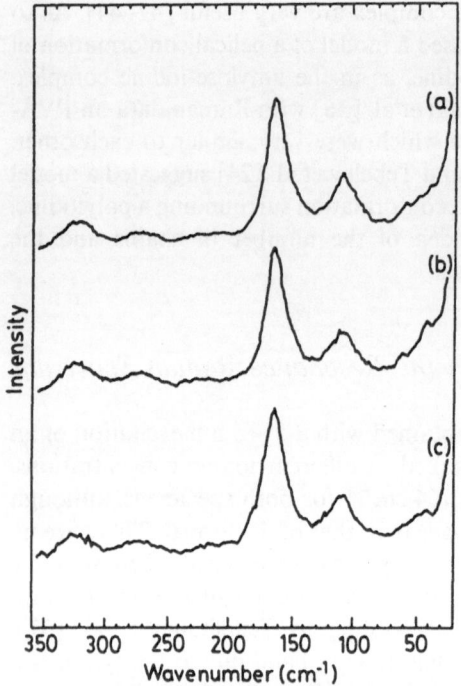

Fig. 16. The resonance Raman spectra (v_0 514.5 nm) of swollen PVA films soaked at various concentrations of iodine solution: (a) 1×10^{-3} mol/l; (b) 2×10^{-3} mol/l; (c) 4×10^{-3} mol/l

polyiodine. These segments form hydrogen bonds between neighbors, and the syndiotactic configuration is necessary for this mode of hydrogen bonding. Each segment has a phase difference along the chain axis of a half of repeat distance (126 pm) so that all the OH groups may form interchain hydrogen bonds whose distance is 307 pm. The condition for the tacticity of the PVA segments may not be very strict, and some of the segments with configurational irregularity may be allowed to participate in the complex formation. This situation is similar to that in the crystallization of PVA: it is a well known fact that atactic PVA, polymerized without any tacticity control, can crystallize with a high degree of crystallinity. However, the condition for the tacticity must be more strict in the complex than in the crystal, for the sequence length of a PVA segment participating in a complex is much shorter than that in a crystal. The irregularity of the tacticity necessarily causes faults in the hydrogen bonding, decreasing the stability of complex. This must be one of the most important reasons why both the capacity for complex formation and the stability of the complex of syndiotactictly-rich PVA are higher than those of atactic PVA. The 1,2 glycol hetero-linkage also causes the faults to decrease the capacity for complex formation of a PVA and the stability of the complex formed as shown by Murahashi [10, 11].

It should be noted that this model was not constructed on direct experimental evidence but on indirect evidence. As for the configuration of PVA, there

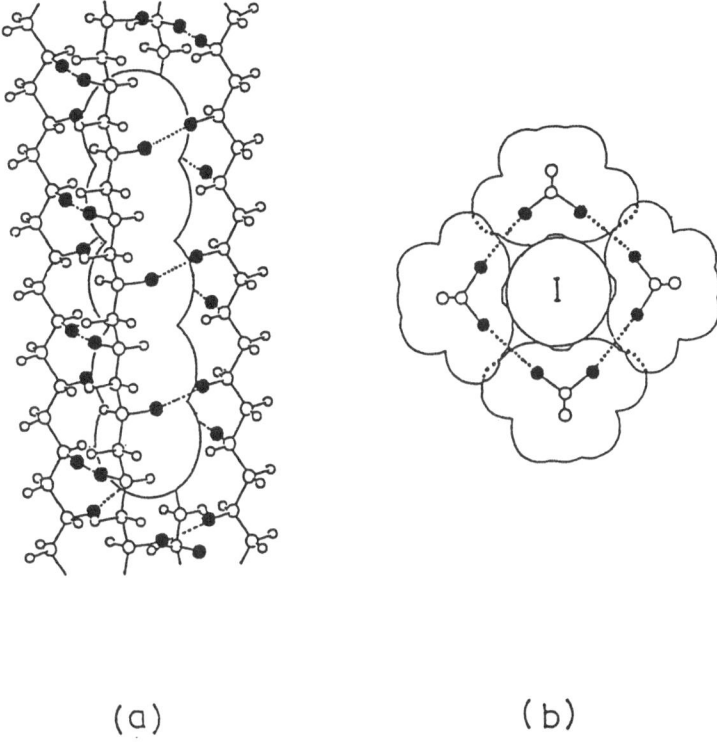

(a) (b)

Fig. 17a, b. Proposed lattice match between I_5- column and PVA chains in the complex: the intermolecular hydrogen bonds are shown by *dashed lines*. On the projection along the complex axis (b), *solid lines* outline van der Warrs radius of each molecule. ○ : hydrogen, ○ : carbon, ● : OH group

is no doubt about the superiority of syndiotactictly-rich PVA for complex formation over atactic PVA. This is also supported by the fact that isotactictly-rich PVA can not form the complex. Figure 18 [60] shows that a syndiotactictly-rich PVA film has a much higher capacity for complex formation than an atactic PVA film. According to this figure, the amount of the complex forming in the s-rich PVA is larger than that in the a-PVA at a given iodine concentration in the soaking solution. The result obtained with the same experiment as that for Fig. 18 showed that the melting points of the complex in a soaking solution were 85 °C for the s-PVA specimen and 60 °C for the a-PVA specimen, indicating the high stability of the s-PVA complex. It should be noted that it is difficult to prepare s-rich PVA films with degrees of swelling as high as those of a-PVA films, because of the high crystallinity. As shown in Sect. 3, the ability of PVA film to swell with water much affects the capacity for complex formation, and therefore the difficulty in preparing s-rich PVA films with a high degree of swelling becomes a serious problem to overcome from the point of view of polarizer production. The information about the tacticity, however, does not

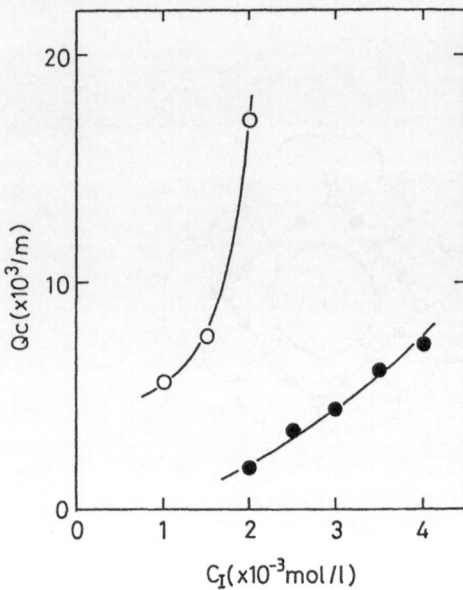

Fig. 18. Relationship between the amount of complex Q_c and the iodine concentration $C_{\bar{I}}$ in soaking solutions.: ○ : s-PVA, ● : a-PVA

give any idea about the choice of the conformation, for the syndiotactic configuration is required, not only for our extended mode structure, but also for Zwick's helical model, and that the isotactic configuration is incompatible with both of them. One of the experimental results supporting the extended conformation is the acceleration of the complex formation by the chain extension discussed in the previous section. Chain extension is doubtlessly favorable for complex formation.

With regard to the number of chains participating in a complex, an interesting result has been obtained on the dichroism of the complex and the dye absorbed in the same PVA film [61]: The degrees of orientation of the complexes formed in extended PVA films are higher than those of dye sorbed in the same film. This implies that not just a single PVA chain but several chains participate in a complex. This is because, if several chains participate in a complex formation, the complex axis tends to orient to the direction of orientation averaged for these chains, which must be the extension direction. On the other hand, in the case of dye, the orientation reflects that of each PVA segment which absorbs the dye, and results in a lower degree of orientation than that of the complex. Figure 19 shows that this expectation is actually satisfied, and supports the idea that several chains must participate in a complex. Finally, the number of chains was assumed to be 4 from the geometrical consideration so that they can cover a polyiodine.

Furthermore, it is reasonable to assume that a tie chain whose freedom of movement is highly restricted by crystal interfaces can hardly form a helix to surround a polyiodine for making a complex. These considerations and experi-

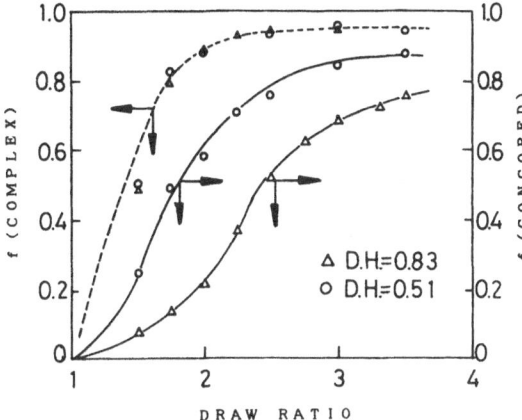

Fig. 19. The degree of orientation of PVA-iodine complex ($f_{complex}$) and congored ($f_{congo\,red}$) as a function of D.H. and draw ratio

mental results led us to make the model of the complex shown in Fig. 17, although they are not direct evidence supporting the model. Zwick proposed his model for the complex forming in the solution, while our model is for the complex formed in water-swollen PVA films. This might be the origin of the different models, for the difference in the freedom of movement of PVA chains is doubtlessly quite large in solutions and films. Nevertheless, we consider that our model characterized by four segments with a planar zigzag conformation must be not only for the complex formed in PVA films but also for that formed in solution. For example, the similarity of the thermal stability of complexes in the solution [26] and in films swollen in the soaking solutions, shown in Fig. 9, is considered to be important evidence that both complexes have the same structure.

5.3 The Coherent Length of Iodines Participating in the Complex

As already cited above, a meridional layer like an X-ray diffraction peak appears due to the polyiodine in drawn PVA films soaked at comparatively low iodine concentration. The breadth of the layer diffraction profile was first analyzed by Haisa and his coworkers [34] and they concluded that a large number of iodine atoms form the lattice: the number of iodine atoms is about three times as large as the 5 for the I_5^- mode complex which is formed predominantly in the amorphous phase. In order to address this problem, we also analyzed in detail the breadth of the meridional layer peak from the iodine lattice with the following results [62]. Figure 20 shows X-ray diffraction photographs of drawn PVA film before and after iodine soaking. Figure 21 shows X-ray meridional intensity profiles of a drawn PVA film measured (a) before and (b) after iodine soaking. The residue (c) after subtraction of the (b) from (a) is considered as showing the contribution of iodine. On the profile (c), a peak at $2\theta = 29°$

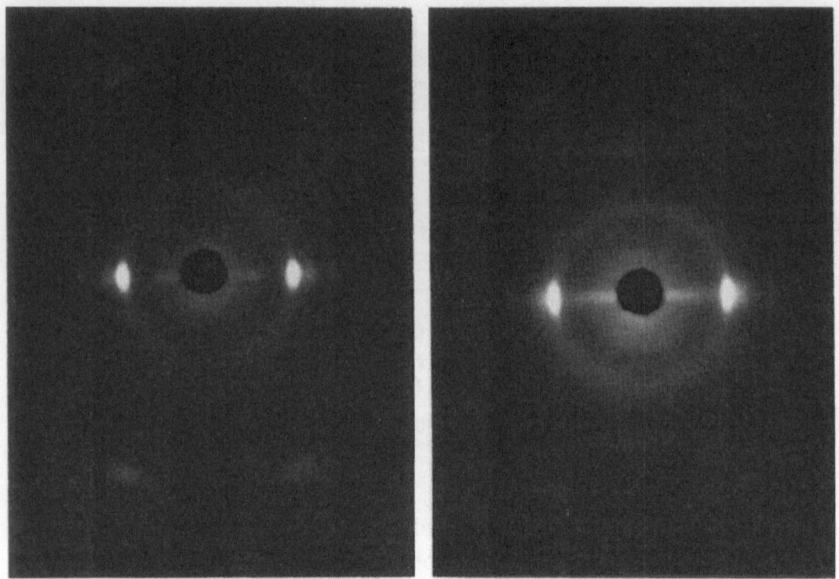

Fig. 20a, b. X-ray diffraction photographs of drawn PVA and iodinated PVA films taken with the incident beam normal to the film plane: (**a**) drawn PVA film ($\lambda = 5$), (**b**) iodinated PVA film soaked at 1×10^{-2} mol/l

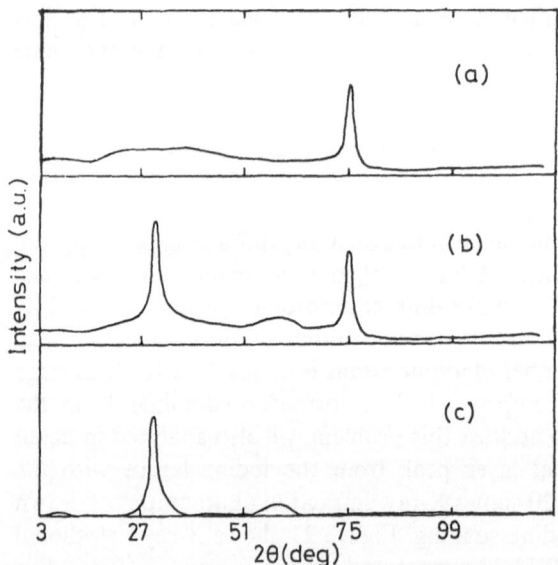

Fig. 21a–c. X-ray meridional intensity profiles: (**a**) drawn PVA film, (**b**) iodinated PVA film, (**c**) iodine ((b)–(a))

corresponds to a spacing of 310 pm, while the peak at $2\theta = 61°$ corresponding to 155 pm is the second order reflection of the former. The sharpness of the X-ray diffraction profile, measured along the scattering angle, depends on the crystallite size and the lattice distortions. According to Hosemann [63], the breadth of a diffraction δ_0 is related to these two factors by Eq (1).

$$\delta_0^2 = (1/L)^2 + (\pi g)^4 \, m^4/d^2 \qquad (4)$$

where L is the length of a lattice, d is the spacing, g is a relative fluctuation of d, and m is the order of the reflection.

The analysis of data on the profile (c) by Eq. (4) gave L = 3.3 nm corresponding a linear lattice consisting of about 11 atoms with a 310 pm repeat distance, and a value of 0.052 for g. This number is about 30% smaller than that of Haisa, but it is certain that the number is still more than twice as large as that of I_5-. It is noted that the evaluated L over 3 nm resulted neither from the improper application of Eq. (4) nor from the error in the evaluation of the breadth. For example, instead of using Eq. (4), a simple Scherrer equation which corresponds to the case of g = 0 gave nearly the same value for L. It is noted that the Scherrer equation gives the smallest value of the crystallite size for a given observed breadth. Furthermore, the Fourier analyses of the intensity profiles (b) and (c) support the results of analyses of these meridional diffraction profiles, showing no discontinuous change which would indicate the existence of the isolated lattice of 5 iodine atoms. Thus there is no doubt that the I_5- mode polyiodine and the I_3- mode, whose fraction is much less than that of I_5-, do not exist separately but coagulate increasing the coherent length for X-rays in the chain direction and resulting in sharper diffraction profiles than those expected from independent polyiodines.

The I_3- mode polyiodines form a super lattice with a spacing of 900 pm within a crystal, demonstrated directly by X-ray diffraction in the specimens soaked at high iodine concentrations, as will be shown in next section. In this case, however, no evidence is obtained for the existence of any super lattice of these polyiodines. Thus the result on the breadth of the profile indicates that the polyiodines coagulate to make an apparent linear lattice in which iodine atoms keep nearly the same distance to their neighbors.

6 Structure and Properties of the Complex Formed in PVA Films Soaked at High Iodine Concentrations

So far we have studied PVA-Iodine complexes which are formed in PVA films soaked at low iodine concentrations at which iodine sorption and complex formation only take place in the amorphous phase. In this case, X-ray diffraction of PVA crystals remained without any change after the iodine soaking, as shown in Fig. 20 in the previous section. Although detailed X-ray measurement of

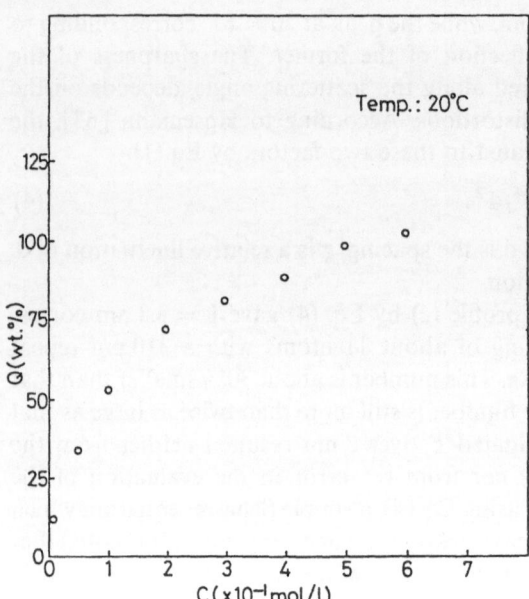

Fig. 22. Sorption iostherm of iodine by PVA films at 20 °C

spacing shows that some of the iodine atoms absorbed penetrate into the crystalline phase, the amount of iodine penetrating into the crystalline phase is considered as being negligibly small. Iodine penetration becomes noticeable when the iodine concentration is over a threshold value as first reported by West [27]. We are interested in what happens in high iodine-concentration soaking since few studies have been made on the subject so far.

In this section, we will discuss PVA-Iodine complexes which are formed at high iodine concentrations of soaking. Figure 22 shows the iodine sorption isotherm of a PVA film, obtained by soaking in iodine-KI solutions at 20 °C. The data in Fig. 22 apparently satisfy a Freundlich relation i.e. lnQ increases linearly with lnC. At 50 wt % sorption, the distribution coefficient is about 6 which is defined as the ratio of iodine concentration per water in a swollen PVA film to that of the soaking solution. This large distribution coefficient is due to the iodine sorption in the crystalline phase as well as in the amorphous phase.

6.1 X-ray Diffraction of Specimens Soaked at High Iodine Concentrations

Figure 23 shows X-ray diffraction photographs of heavily iodinated drawn films. The diffraction photographs of the original and lightly iodinated specimens have already been shown in Fig. 20. New equatorial spots begin to appear when the iodine concentration is more than 1.5×10^{-2} mol/l, resulting in an iodine

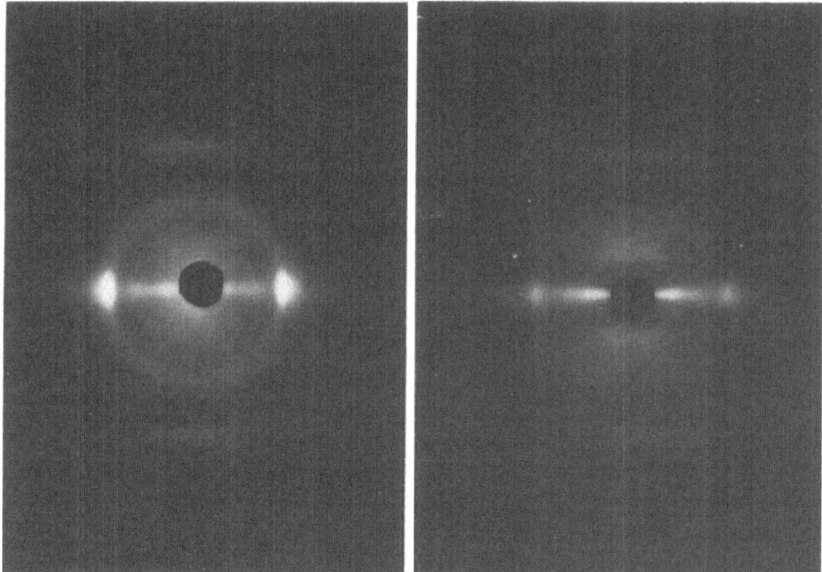

Fig. 23a, b. X-ray diffraction photographs of iodinated PVA films taken with the incident beam normal to the film plane: (a) 2×10^{-2} mol/l, (b) 4×10^{-1} mol/l

sorption over 12 wt %, as already reported by Hess [48]. As mentioned above, a slight change of PVA crystal spacing is induced even by soaking at low iodine concentrations.

Figure 24 shows the equatorial diffraction intensity curves of highly iodinated PVA specimens. In the specimen soaked at 2×10^{-2} mol/l (14 wt % sorption), the peaks at $2\theta = 6.6°$ and $13.5°$ corresponding to a spacing of 1338 pm and 658 pm are not related to the PVA crystal. At the same time, the peak at $19.5°$ which is originally assigned to the $(1\,0\,1)$ and $(1\,0\,\bar{1})$ doublet of the PVA crystal, shifts to a higher angle with increasing iodine concentration. These results prove that some PVA-Iodine co-crystals are formed in heavily iodinated samples, although the cells are not uniform, as suggested by extraordinary broadening of the diffraction profiles. In spite of remarkable changes in the equatorial diffractions, the $(0\,2\,0)$ diffraction characteristic to the PVA crystal still remains as it was before iodine sorption, even at 50 wt % iodine sorption. This means that the periodicity is much less affected by iodine sorption in the chain direction than in the perpendicular direction. On the basis of these equatorial X-ray diffractions, we propose the following model for the co-crystal:

Figure 25 shows the projection of the unit cell onto the a–c plane perpendicular to the chain axis (b axis). Here 215 pm was taken as the van der Waals radius of an iodine atom. The cell is approximately orthorhombic. The spacings shown in Table 1 correspond to the intensity curve (d) in Fig. 24. This model is

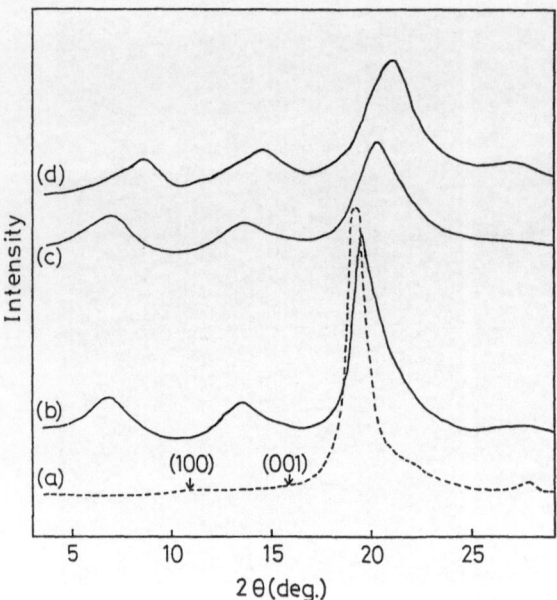

Fig. 24. Change of equatorial X-ray diffraction pattern of iodinated PVA films with iodine sorption: Number in a mol/l unit correspond to the iodine concentration in soaking solution: (a) 0 mol/l, (b) 2×10^{-2} mol/l, (c) 1×10^{-1} mol/l, (d) 4×10^{-1} mol/l

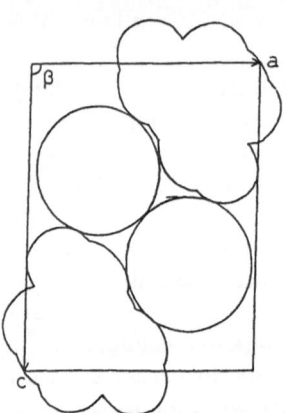

Fig. 25. Substitution model of the unit cell of "iodine-PVA" crystal in the a–c plane

constructed on the basis of Bunn's PVA crystal structure in which two chains of the syndiotactic configuration are arranged so that all OH groups may form inter-chain hydrogen bonds: in this cell one of the two PVA chains in Bunn's cell is substituted by a polyiodine, followed by some slight adjustment, and therefore the c axis is about twice as long as that of Bunn's PVA cell. Table 1 shows comparatively good agreement between observation and calculation. We also

Table 1. Comparison of observed and calculated spacings and intensities of an iodinated PVA film

(h k l)	d-spacing (Å)		Intensity[a]	
	Observed	Calculated	Observed	Calculated
0 0 1	10.04	10.0	s	s
1 0 0	7.6	7.8	shoulder	w
1 0 $\bar{1}$	6.07	{ 6.2	s	s
1 0 1		6.03		
0 0 2	5.0	5.0	shoulder	w
1 0 2	4.23	{ 4.15	vs	vs
$\bar{1}$ 0 2		4.26		
0 0 3	3.29	3.33	w	w

[a] vs = very strong, s = strong, w = weak.

Fig. 26. Intercalation model of the unit cell of "iodine-PVA" crystal in the a–c plane

consider that another cell shown in Fig. 26 might be formed as well as that of Fig. 25. However, it should be noted that the model in Fig. 26 identifies the peaks at $2\theta = 6.5°$ and $19.4°$ well, but fails to do so for peak at $13.5°$.

The continuous change of the diffraction and extraordinary broadening of the profile indicate that the intrusion of iodine does not occur uniformly through a whole crystallite resulting in many kinds of co-crystal lattices. Under these circumstances, it should be recognized that the models proposed here are idealized ones. On the other hand, this situation must introduce a large strain into the crystal, as suggested by the broadening of diffraction profiles. This makes us expect that the desorption of iodine must occur very easily and quickly from the crystalline phase. This expectation is well satisfied as will be shown later.

In the highly iodinated specimens, a new meridional peak appears at about $2\theta = 28°$, indicating the origination of a new mode of polyiodine with an inter-atom period of 320 pm. As shown in the proceeding section, a peak at $2\theta = 29°$, corresponding to a 308 pm spacing, appears in samples soaked at iodine concentrations less than 1.5×10^{-2} mol/l, which is due to the complex forming in the amorphous phase. Iodine is absorbed in the crystalline phase as well as in the amorphous phase in highly iodinated specimens, and the iodine absorbed in the crystalline phase forms some co-crystals whose structures are proposed above. At the same time, the absorption spectrum of the heavily iodinated specimen shows that iodine in the crystal forms the I_3- mode complex, as will be shown shortly. Thus the spacing of 320 pm is identified to the polyiodine forming the co-crystals.

In addition to these peaks with spacings of about 320 pm and 308 pm, new meridional peaks appear at $2\theta = 9.2°$ and $2\theta = 18.5°$, corresponding to 960 pm and 480 pm, respectively. The latter peak is considered to be the second order diffraction of the former. This appearance of new peaks indicates that I_3- mode polyiodines form a linear super lattice in the crystal phase. Herbstein [64] et al. studied triodides with well established structures to show that the repeat distance of linear lattice of I_3- mode polyiodines is between 959 and 989 pm, and proposed that the shorter repeat distance is caused by kinking of successive I_3- ions. Thus the 960 pm repeat distance observed here may imply that I_3- mode polyiodines form a linear lattice within a crystal, accompanying a kinking. If this kinking of I_3- mode polyiodines is really caused, some slight revision may be required for the models in Figs. 25 and 26.

The structure of co-crystals proposed here necessarily causes breakage of inter-chain hydrogen bonds between OH groups. Figure 27 shows IR spectra

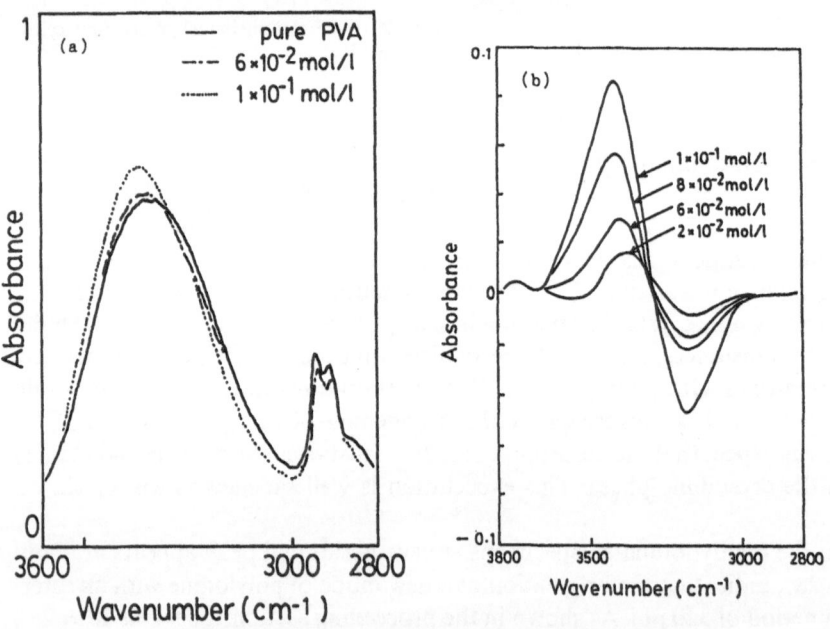

Fig. 27a. Infrared and (**b**) difference spectra of the iodine-sorbed PVA films soaked in solution at various iodine concentrations

and difference spectra in a region from 2800 to 3800 cm^{-1} of the specimens soaked at different iodine concentrations. The broad band slightly shifts to the higher frequency region with increasing iodine concentration. This shift is due to a slight decrease in absorbance in the lower frequency region and a slight increase in the higher frequency region. It is well known that the lower frequency band is related to OH groups forming hydrogen bonds, and the higher side one is related to unbonded OH groups.

6.2 Resonance Raman Spectra of Specimens Soaked at High Iodine Concentrations

Figures 28 and 29 show resonance Raman spectra of two specimens soaked at low and high iodine concentrations. In the previous section, both the 109 cm^{-1} and the 161 cm^{-1} peak are assigned to I_3- and I_5- mode polyiodines, respectively. According to Fig. 28 the intensity ratios are different between the two specimens: the 109 cm^{-1} peak is stronger than the 161 cm^{-1} peak in the heavily iodinated specimen. This corresponds to the fact observed at high iodine concentrations that the I_3- mode polyiodines is formed within crystals and that the visible light absorption peak due to I_3- mode complex remarkably increases its intensity as shown by Fig. 29. According to Fig. 29, the 109 cm^{-1} peak intensity is enhanced much more by 488 nm excitation than by 514.5 nm excitation.

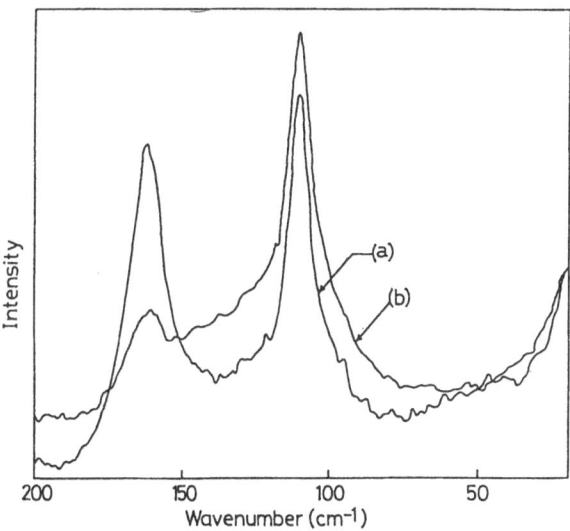

Fig. 28. Resonance Raman spectra (v_0 514.5 nm) of iodinated PVA films soaked in iodine solutions of different iodine concentrations. (**a**) 5×10^{-3} mol/l, (**b**) 1×10^{-1} mol/l

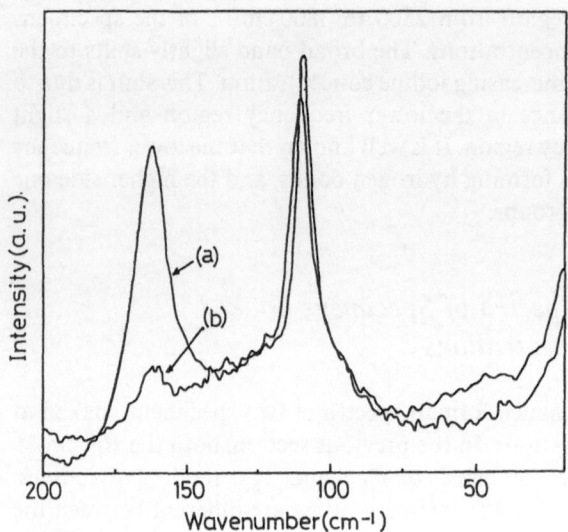

Fig. 29. Exciting frequency (v_0) dependence of the resonance Raman spectra of a iodinated PVA films soaked in a iodine solution (5×10^{-3} mol/l). (*a*) 514.5 nm, (*b*) 488.0 nm

6.3 Absorption Spectra of Specimens Soaked at High Iodine Concentrations

Figure 30 shows the visible light absorption spectra of an annealed thin film soaked at different iodine concentrations. The absorption peak at 350 nm is due to the iodine-KI solution, but not to any complex. The peak at 580 nm in curve (d) obtained at low concentrations is mainly caused by the I_5- mode complex, forming mostly in the amorphous phase as mentioned above. The absorption peak at about 470 nm in curve (a), obtained at high iodine concentrations, is caused by the I_3- mode complex. It should be noted that the 470 nm peak increases in intensity remarkably. In these specimens, iodine sorption takes place not only in the amorphous phase but also in the crystalline phase. The change of the spectrum from (a) to (d) shows that the amounts of both I_5- and I_3- modes of the complex increase with increasing iodine concentration, and that the fraction of the I_3- mode shows a remarkable increase: The I_5- mode complex prevails over that of the I_3- mode at low iodine concentrations, while the amount of I_3- mode complex is predominant in the crystal at high iodine concentrations. These results show that at high iodine concentrations, iodine penetrating into the crystalline phase forms the I_3- mode complex within crystals, while resulting in remarkably disordered co-crystals, as discussed above.

Figure 31 shows the change of the absorption spectrum with desorption of iodine: a thin PVA film was first soaked in a 1×10^{-1} mol/l solution, corres-

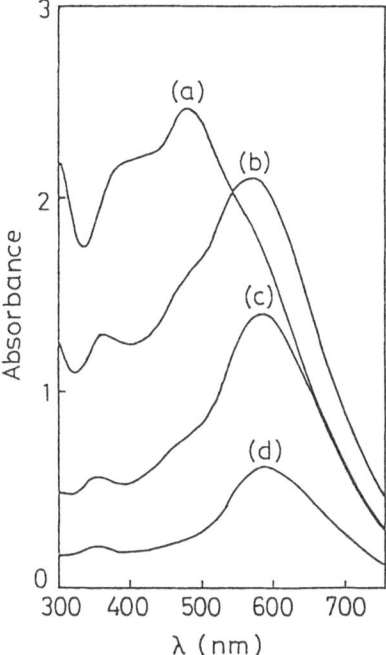

Fig. 30. Absorption spectra of iodinated as-cast PVA films soaked at different iodine concentrations. (a) 1×10^{-1} mol/l, (b) 5×10^{-2} mol/l, (c) 2×10^{-2} mol/l, (d) 5×10^{-3} mol/l at 20 °C

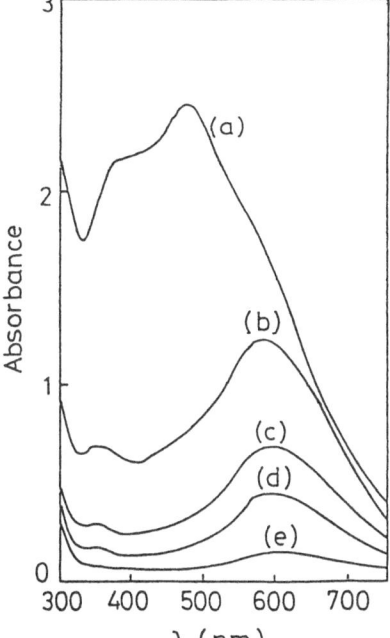

Fig. 31. Charges of absorption spectra with iodine desorption for iodinated as-cast PVA films soaked at 1×10^{-1} mol/l iodine concentration. (a) 0 s, (b) 1 s, (c) 5 s, (d) 10 s, (e) 30 s.

ponding to the curve (a) in Fig. 30, and then dipped in pure water at 20 °C for a given length of time. The 470 nm peak very quickly disappears, while the 580 nm peak gradually decreases in intensity. It should be noted that the length of time indicated for each curve in Fig. 31 is the time of water dipping not the real desorption time, because the desorption must have continued even after the film had been removed from water, particularly in the crystalline phase. The iodine leaving the crystalline phase moves into the amorphous phase. This must be one of the reasons for the rapid disappearance of the 470 nm peak and the slow decrease of the 580 nm peak during desorption. However, the low stability of the I_3- mode complex formed in the crystalline phase, giving the 470 nm absorption peak must be the most important reason for the rapid disappearance. It is interesting to compare the iodine sorption behavior of PVA with that of some other polymers. In nylon-6 [65] and polyacetylene [66], iodine sorption in the crystalline phase takes place when the activity of iodine in the solution and vapor reaches a distinct threshold. This is similar to the case of PVA, although thresholds in those polymers are clearer than in the case of PVA. However, the desorption from the crystalline phase in nylon-6 and polyacetyline is quite different from that in PVA: the desorption of iodine from nylon-6 crystals is not caused by water but by an aqueous $Na_2S_2O_3$ solution, which shows a high capacity for complex formation in the nylon-6 crystal phase as well as in the amorphous phase. This quick desorption of iodine from the PVA crystal phase must be related to the disordered lattice structure of the co-crystals, as mentioned above.

7 Conclusion

Our structure studies on the PVA-Iodine complexes formed in PVA films soaked in iodine-KI solutions without Boric acid have been reviewed in this paper.

In the first section, the studies on the complex which have been made by many authors from various points of view were reviewed. Most of the studies on the complex have been made in aqueous PVA-Iodine-KI solutions with Boric acid. Our group, however, has been interested in the structural studies of the complex formed in PVA films. In Sect. 1, the importance of the structural study on the complex, as well as the problems and the intention of our studies were described.

In Sect. 2, a double network structure was proposed for the PVA films in the water-swollen state, i.e. an amorphous chain (tie chain) network and a fibrillar network. The blue color complex is considered to be formed mostly in the former network. Subsequently what happens in PVA films during soaking in aqueous solutions were discussed. The contractions of the volume and the long spacing from SAXS were compared, and the difference between the contractions was discussed on the basis of the double network model of water-swollen PVA films.

In Sect. 3, the effects of the equilibrium degree of swelling of PVA films, depending on the annealing temperature, on the formation and the properties of the complex were considered as functions of soaking temperature and iodine concentration. The equilibrium degree of hydration which was taken as the parameter characterizing the fine structure of PVA films greatly affects both the formation and the properties of the complex: the larger the degree of hydration, the greater the amount of complex formed and the larger the λ_{max}. However, the thermal stability of the complex measured in the soaking solution is lower in the specimen with a larger degree of hydration. It was found that the thermal stability of the complex is closely related to the thermal motion of water swollen PVA chains which has a transition point at 20 °C.

In Sect. 4, the effect of the extension of the amorphous chains on complex formation was studied. Specimens were stretched in the soaking solutions whose iodine concentrations are too low for the complex formation. It was found that the complex begins to form during the extension, and that although the extension energy required for the complex formation depends to a great extent on the degree of hydration, the dependence almost disappears when standardized by the weight of PVA. The enhancement of the complex formation by chain extension implies that the chain conformation of PVA may not be helical but rather an extended mode in the complex.

In Sect. 5, the structure of the complex was studied by X-ray diffraction, resonance Raman spectrum and visible ray absorption, in a comparatively low range of iodine concentration where iodine sorption takes place mostly in the amorphous phase of PVA. The I_5- mode complex is formed predominantly in this case, although the I_3- mode complex is also formed. However, the breadth of the X-ray diffraction profile due to the polyiodine indicates that the polyiodines coagulate to make an iodine column almost 3 times as long as that of I_5- mode polyiodine. A structural model for the complex forming in the amorphous phase was proposed, and it is characterized by four syndiotactic PVA chains of planar zigzag conformation, surrounding a polyiodine. These four chains are fixed with interchain hydrogen bonds.

In Sect. 6, the complex formation in the crystalline phase of PVA which is caused at very high iodine concentrations in the soaking solution was studied. In this case, the X-ray diffraction completely changes from that observed at low iodine concentrations, and both resonance Raman and visible light absorption spectra demonstrate that the I_3- mode complex is formed in the crystalline phase. The iodine absorbed in the crystalline phase forms a co-crystal with PVA whose model is proposed on the basis of the equatorial X-ray diffraction. However, it should be noted that the adsorption does not occur uniformly throughout a crystal, as suggested by the extraordinary broadening of the X-ray diffraction. The structure of the complex formed in the crystalline phase is reflected by the very quick desorption of iodine from the crystalline phase. X-ray diffraction shows that the I_3- mode polyiodines form a linear super lattice with some kinking in the crystalline phase.

8 References

1. Colin, Gaultier de Clouby H (1814) Ann Chim 90: 87; (1814) Gilb Ann 48: 297
2. Freudenberg K, Schaaf E, Dumpert G, Ploetz T (1939) Naturwissenschaften 22: 850
3. Arimoto H (1962) Kobunshi Kagaku 19: 101
4. Scholtan W (1954) Makromol Chem 11: 131
5. Herrmen WO, Haehnel W (1927) Ber Dtsch Chem Ges 60: 1658
6. Staudinger H, Frey K, Starck W (1927) Ber Dtsch Chem Ges 60: 1782
7. Land EH (1951) J Opt Soc Am 41: 957
8. Tanizaki Y (1957) Bull Chem Soc Japan 30: 935
9. Munakata H, Ichikawa R (1978) Sen-i Gakkaishi 34: 288; Moriuchi T (1984) Kobunshi 33: 830
10. Kikukawa K, Nozakura S, Murahashi S (1971) Polym J 2: 212
11. Kikukawa K, Nozàkura S, Murahashi S (1971) Polym J 3: 52
12. Patent publication 206402-91 issued in Japan (Sep. 9. 1991)
13. Imai K, Matsumoto M (1961) J Polym Sci 55: 335
14. Matsuzawa S, Yamaura K, Noguchi H (1974) Makromol Chem 175: 31
15. Choi YS, Miyasaka K (submitted to J Appl Polym Sci)
16. Hayashi S, Nakano C, Motoyama T (1963) Kobunshi Kagaku 20: 303
17. Hayashi S, Tanabe Y, Hojo N (1977) Makromol Chem 178: 1679
18. Hayashi S, Kabayashi M, Shirai H, Hojo N (1978) Makromol Chem 179: 1397
19. Hayashi S, Takayama M, Kawamura C (1970) Kogyo Kagaku Zasshi 73: 178
20. Hayashi S, Takayama M, Kawamura C (1970) Kogyo Kagaku Zasshi 73: 412
21. Sakuramachi H, Choi YS, Miyasaka K (1990) Polym J 22: 638
22. Gallay W (1936) Can J Res 14B: 105
23. Zwick MM (1965) J Appl Polym Sci 9: 2393
24. Tebelev LG, Milkulskii GF, Korchagina YP, Glikman SA (1965) Vysokomol Soedin 7: 1231
25. Oishi Y, Miyasaka K (1986) Polym J 18: 307
26. Yokota T, Kimura Y (1984) Makromol Chem 185: 749
27. West CD (1949) J Chem Phys 17: 219
28. Saito S, Okutama H, Kishimoto H, Fujiyama T (1955) Kolloid ZZ Polym 144: 41
29. Pritchard JG, Akintola DA (1972) Talanta 19: 877
30. Deuel H, Neukom H (1949) Makromol Chem 3: 13
31. Shibayama M, Sato M, Kimura Y, Fujiwara H, Nomura S (1988) Polymer 29: 336
32. Voelkel J, Szydlowska W (1981) Makromol Chem 182: 225
33. West CD (1947) J Chem Phys 15: 689; (1951) Makromol Chem 19: 1432
34. Haisa M, Itami H (1957) J Phys Chem 61: 817
35. Heyde ME, Rimai L, Kilponen RG, Gill D (1972) J Am Chem Soc 94: 5222
36. Inagaki F, Harada I, Shimanouchi T, Tasumi M (1972) Bull Chem Soc Jpn 45: 3384
37. Teitelbaum RC, Ruby SL, Marks TJ (1980) J Am Chem Soc 102: 3322
38. Rundle RE, Foster JF, Baldwin RR (1944) J Am Chem Soc 66: 2116; Rundle RE, Baldvoin RR (1943) J Am Chem Soc 65: 554; Rundle RE, French D (1943) J Am Chem Soc 65: 558
39. Handa T, Yajima H (1979) Biopolymers 18: 873; (1980) 19: 723; (1980) 19: 1723
40. Saenger W (1984) Naturwissenschaften 71: 31
41. Zwick MM (1966) J Polym Sci Part A-1 4: 1642
42. Yokota T, Kimura Y (1985) Makromol Chem 186: 549
43. Yokota T, Kimura Y (1986) Makromol Chem Rapid Comm 7: 249
44. Yokota T, Kimura Y (1989) Makromol Chem 190: 939
45. Hayashi S, Takizawa K (1968) Kogyo Kagaku Zasshi 71: 101
46. Kojima Y, Furuhata K, Miyasaka K (1985) J Appl Polym Sci 30: 1617

47. Oishi Y, Miyasaka K (1987) Polym J 19: 331
48. Hess VK, Steinman R, Kiessig H, Avisiers I (1957) Kolloid Z 153: 128
49. Kojima Y, Furuhata K, Miyasaka K (1983) J Appl Polym Sci 28: 2401
50. Oishi Y, Yamamoto H, Miyasaka K (1987) Polym J 19: 1261
51. Choi YS, Oishi Y, Miyasaka K (1990) Polym J 22: 601
52. Abitz W, Gerngross O, Herrmann K (1930) Naturwissensshaften 18: 754; (1903) Biochem 228: 409
53. Hearle JWS (1968) Fiber structure, Butterworth, p 214
54. Keller A (1957) Phil Mag 2: 1171
55. Hosemann R (1962) Polymer 3: 349
56. Flory PJ (1953) Principles of polymer chemistry. Cornel University Press, Ithaca, New York, chap 13
57. Tanaka M, Mizutani K (1979) Angew Makromol Chem 78: 211
58. Hayashi S, Kawamura C, Takayama T (1970) Bull Chem Soc Jpn 43: 537
59. Sarles LR, Cottos RM (1958) Phys Rev 111: 853
60. Choi YS, Miyasaka K (1900) Polym Preprints Jpn 39: 2436
61. Kim KH, Ieda Y, Miyasaka K (Submitted to Polym J)
62. Choi YS, Miyasaka K (1991) Polym J 23: 977
63. R Hosemann, Bagchi SN (1962) Direct analysis of diffraction by matter. North Holland, Amsterdam, p 239
64. Herbstein FH, Kaftory M, Kapon M, Saenger W (1980) Z Kristallogr 154: 11
65. Ueda S, Kimura T (1958) Kobunshi Kagaku 15: 243
66. Danno T, Miyasaka K, Ishikawa K (1983) J Polym Sci, Phys Edn 21: 1527

Received January 22, 1992

Author Index Volume 101–108

Subject Index